SpringerBriefs in Electrical and Computer Engineering

More information about this series at http://www.springer.com/series/10059

Lei Lei • Chuang Lin • Zhangdui Zhong

Stochastic Petri Nets
for Wireless Networks

 Springer

Lei Lei
State Key Laboratory of Rail
 Traffic Control and Safety
Beijing Jiaotong University
Beijing, China

Chuang Lin
Department of Computer Science
 and Technology
Tsinghua University
Beijing, China

Zhangdui Zhong
School of Computer
 and Information Technology
Beijing Jiaotong University
Beijing, China

ISSN 2191-8112   ISSN 2191-8120   (electronic)
SpringerBriefs in Electrical and Computer Engineering
ISBN 978-3-319-16882-1   ISBN 978-3-319-16883-8   (eBook)
DOI 10.1007/978-3-319-16883-8

Library of Congress Control Number: 2015937432

Springer Cham Heidelberg New York Dordrecht London

Printed on acid-free paper

Springer International Publishing AG Switzerland is part of Springer Science+Business Media (www.springer.com)

Recommended by Xuemin (Sherman) Shen

Preface

Stochastic Petri Nets (SPNs), introduced in 1980, are a modeling formalism that can be conveniently used for the performance and reliability evaluation of discrete event systems. They admit a graphical representation that is well suited to top-down and bottom-up modeling of complex systems, and present a very straightforward mapping between events in the SPN model and events in the underlying Markov process. Although SPNs have become a useful tool for researchers in computer science, they are unknown to most wireless researchers and are not widely used to model wireless communication systems. On the other hand, the next-generation wireless networks such as the 5th Generation (5G) cellular systems will become increasingly complex in order to support for an increasingly diverse set of services, applications, and users—all with extremely diverging performance requirements. Since SPNs are found to be powerful in modeling performance of computer systems with a wealth of numerical solution techniques, it is very interesting to explore their applicability in wireless systems. This book was motivated by a desire to bridge the gap between the research on SPN modeling formalism and on the performance modeling of wireless networks.

In this book, we present our research results on applying SPNs to the performance evaluation of wireless networks under bursty traffic, in terms of typical Quality-of-Service (QoS) performance metrics such as mean throughput, average delay, packet dropping probability, etc. In the first chapter, we introduce the key motivations, challenges, and state-of-the-art research on using SPNs for cross-layer performance analysis in wireless networks. In Chap. 2, we first introduce the SPN basics, and then focus on two powerful techniques in SPNs to deal with the well-known state space explosion problem: (1) model decomposition and iteration; (2) model aggregation using Stochastic High-Level Petri Nets (SHLPNs). We apply the first technique to the performance analysis of opportunistic scheduling and Device-to-Device (D2D) communications with full frequency reuse between D2D links in Chaps. 3 and 4, respectively. The above two scenarios show two typical radio resource sharing paradigms in wireless networks: orthogonal sharing by scheduling and non-orthogonal sharing by frequency reuse. We show that SPNs can provide an intuitive and efficient way in modeling the multiuser wireless system, especially facilitating the inclusion of different resource sharing paradigms between wireless links. Moreover, the original complex model whose state space grows exponentially with the number of users can be decomposed into multiple single user subsystems, and iteration methods can be used for performance approximation. In Chap. 5, we apply the second technique to formulate a wireless channel model for Orthogonal Frequency Division Multiplexing (OFDM) multi-carrier systems with SHLPN formalism in order to simplify the cross-layer performance analysis of modern wireless systems. Compared with existing Finite State Markov Channel (FSMC) model whose state space grows exponentially with the number of OFDM subchannels, our proposed SHLPN model uses state aggregation technique to deal with this problem. Closed-form expressions to calculate the transition probabilities among the compound markings of the SHLPN model are provided. When applied to derive the performance measures for OFDM system, the SHLPN model can accurately capture the correlated time-varying nature of wireless channels. We believe the example applications of SPNs to wireless networks and related findings will reveal useful insights for the design of radio resource management algorithms and spur a new line of thinking for the performance evaluation of future wireless networks.

Beijing, China   Lei Lei
                                                                                              Chuang Lin
                                                                                           Zhangdui Zhong

Acknowledgements

The authors would like to acknowledge the support of the NSFC (Projects No. 61272168, No. U1334202, and No. 61472199), the State Key Laboratory of Rail Traffic Control and Safety in Beijing Jiaotong University (No. RCS2014ZT10), and the Key Grant Project of Chinese Ministry of Education (No. 313006). Thanks also are due to the Master candidates: Miss Qingyun Hao and Mr. Huijian Wang for their contribution of editing work.

A very special thanks to Prof. Xuemin (Sherman) Shen, the SpringerBriefs Series Editor on Wireless Communications. This book would not be possible without his kind support during the process. Thanks also to the Springer Editors and Staff, all of whom have been exceedingly helpful throughout the production of this book.

Contents

Acronym

3G	3rd-Generation
3GPP	3rd Generation Partnership Project
4G	4th Generation
5G	5th Generation
AMC	Adaptive Modulation and Coding
BS	Base Station
CA	Channel-Aware
CAC	Call Admission Control
CDMA	Code Division Multiple Access
CPN	Colored Petri Net
CQA	Channel/Queue-Aware
CR	Cognitive Radio
CTMC	Continuous-Time Markov Chain
D2D	Device-to-Device
DCA	Dynamic Channel Allocation
DCF	Distributed Coordination Function
DEDS	Discrete Event Dynamic Systems
DSPN	Deterministic and Stochastic Petri Net
DTMC	Discrete-Time Markov Chain
DTSPN	Discrete Time Stochastic Petri Net
FSMC	Finite State Markov Channel
FR	Full Reuse
GE	Gilbert-Elliot
GSPN	Generalized Stochastic Petri Net
HLPN	High-Level Petri Net
HSPDA	High Speed Downlink Packet Access
ISI	Inter-Symbol Interference
LCR	Level-Crossing Rate
LTE	Long Term Evolution
M2M	Machine-to-Machine
MAC	Medium Access Control

MC	Markov Chain
MDP	Markov Decision Process
MDPN	Markov Decision Petri Net
MMDP	Markov Modulated Deterministic Process
MIMO	Multiple-Input Multiple-Output
M-QAM	M-ary Quadrature Amplitude Modulation
NRT	Non-Realtime
OFDM	Orthogonal Frequency Division Multiplexing
OFDMA	Orthogonal Frequency Division Multiple Access
OS	Opportunistic Scheduling
PF	Proportional Fair
PN	Petri Net
QA	Queue-Aware
QoS	Quality-of-Service
RR	Round-Robin
RT	Realtime
SHLPN	Stochastic High-Level Petri Net
SINR	Signal to Interference and Noise Ratio
SNR	Signal to Noise Ratio
SPN	Stochastic Petri Net
SRN	Stochastic Reward Net
SWN	Stochastic Well-Formed Petri Net
TDMA	Time Division Multiple Access
UE	User Equipment
UMTS	Universal Mobile Telecommunications System

Chapter 1
Introduction

Compared with other high-level modeling formalisms such as queuing theory, SPNs have received relatively little attention in the research of wireless networks. This is mostly due to the fact that SPNs are unknown to most wireless researchers. On the other hand, SPNs are a well-developed theory and have been actively studied in the computer science field for the last 20 years. They present a very straightforward mapping between events in the SPN model and events in the underlying Markov process. Most importantly, a set of powerful techniques have been developed in dealing with the well-known state space explosion problem in performance evaluation. In this chapter, we first briefly introduce the general background of cross-layer performance analysis of wireless networks using stochastic models. Then, we focus on the motivations and challenges on using SPNs for performance evaluation. Finally, we provide an overview of the state-of-art research on applying SPNs to the performance evaluation of wireless networks.

1.1 Cross-Layer Performance Analysis of Wireless Networks Using Stochastic Models

The next-generation wireless networks such as the 4th Generation (4G) and 5G cellular systems are targeted at supporting various applications such as voice, data, and multimedia over packet-switched networks. The performance of such networks is evaluated in terms of QoS contract satisfaction under diverse traffic conditions. Performance evaluation can be performed by system-level simulation, where all effects and algorithms of every network layer are implemented in software. Although the extracted performance is accurate, the simulation-based approach is usually time-consuming and cannot be directly applied to the design of optimal mechanisms and algorithms. Therefore, performance evaluation by high-level modeling formalism, e.g., queuing networks or SPN, is an attractive

© The Author(s) 2015
L. Lei et al., *Stochastic Petri Nets for Wireless Networks*, SpringerBriefs
in Electrical and Computer Engineering, DOI 10.1007/978-3-319-16883-8_1

alternative. When formal methods are applied to performance evaluation, stochastic models, especially Markovian models are formulated to characterize the dynamic behavior of the wireless systems. Such a performance model is usually considered as "cross-layer" if it not only characterizes the physical layer aspects but also the behavior of the higher layers, e.g., Medium Access Control (MAC) layer. The key performance measures can be identified and analyzed, revealing the relationships between them if necessary. Moreover, the optimal system structure and operating mode can be designed based on the stochastic models, where the system fulfills all requirements concerning QoS as well as all technical and economic constraints with the given workload.

There are several reasons why performance evaluation of wireless networks need to be based on stochastic models instead of the much simpler deterministic models:

1. *Stochastic nature of wireless channel conditions*: Compared with wireline networks, the performance of wireless systems is dominated by the channel between antennas. The wireless channel conditions vary in both the time domain and frequency domain due to complex phenomena such as multipath fading, Doppler, and time-dispersive effects introduced by the wireless propagation. Therefore, it is critical for networking researchers and engineers to capture the stochastic channel characteristics in their cross-layer performance model.

2. *Stochastic nature of traffic arrivals*: Due to the complex wireless channel characteristics, the stochastic nature of traffic arrivals is usually neglected for simplification when studying the performance of wireless networks, especially when only physical layer performance is considered. The infinitely backlogged traffic model is used where each user always has data to transmit. However, in practice, data arrival process at the users is dynamic and bursty. Therefore, with the accelerating growth of mobile internet and other new wireless applications, more and more research work focuses on bursty traffic models.

3. *Stochastic nature of underlying geometry*: User mobility is one salient feature of wireless networks, which leads to the stochastic nature of underlying geometry (the relative locations of nodes). Since the underlying geometry of wireless networks plays a fundamental role due to the interference of other transmitters, it needs to be characterized by related performance models to analyze key performance metrics, e.g., Signal to Interference and Noise Ratio (SINR).

Performance models for wireless networks can be generally classified into two broad categories. The first category focuses on the investigation of performance at the packet level with an assumption of a static user population [1–7]. The time scale of the packet-level model is the frequency of scheduling algorithms. The service rates of the users are the same as the instantaneous channel transmission rates, which vary randomly over time due to channel fluctuation. The traffic pattern is usually assumed to be saturated with infinite backlogs (i.e., each user always has data to transmit) or features dynamic packet arrivals [8]. For the saturated model, a common objective is to optimize some utility functions of the throughput; while for dynamic packet arrivals, the focus is on network stability (i.e., the queue occupancy can be bounded whenever feasible), the statistical worst case performance (e.g., the tail distribution of packet delay) and average performance (e.g., the average delay, the packet dropping probability). The second category investigates performance at the flow level with time-variant user population [9–11]. In the flow-level analysis, new users arrive according to a stochastic process, and each user has a finite-length file for transmission. A user leaves the system when the entire file is transmitted. The time scale of flow level model is the frequency of the user or flow arrival and departure. Since the full analysis of flow-level performance can be very complicated, a simple constant-rate service process (e.g., the time average value of the channel rate) is usually used to approximate the time-varying channel transmission rate. Important flow-level performance metrics include the distribution of the number of flows, flow throughput and mean response time. Compared to the packet-level analysis, the flow-level analysis is based on more practical traffic patterns, which consider the dependence of performance on user population [9]. However, compared to the flow level analysis, the packet level analysis shows advantages in involving more realistic channel models, which include the simplest memoryless on-off channel, the two-state Gilbert-Elliot channel, and the more complicated and accurate FSMC.

There are several high-level formalisms which provide powerful model formulation and solution techniques for the performance evaluation of wireless networks, e.g., queuing theory, SPNs, stochastic network calculus, and stochastic geometry, etc. In this book, we will focus on the SPNs and discuss its applications to the performance evaluation of wireless networks.

1.2 Motivations and Challenges on Using SPNs for Performance Evaluation

SPNs were introduced in 1980 as a formalism for the description of Discrete Event Dynamic Systems (DEDS) whose dynamic behavior could be represented by means of continuous-time homogeneous Markov chains. They are the stochastic extension of Petri Nets (PNs), which are a powerful tool for the description and the analysis of systems that exhibit concurrency, synchronization and conflicts. Compared with PNs, random variables have been added in SPNs to represent the duration of activities, or the delay until events. SPNs present a very straightforward mapping between events in the SPN model and events in the underlying Markov process. They maintain a clearly arranged graphical structure, and it is easy to generate the Markov process from any SPN model. Moreover, SPNs have the ability to structure complex models into modules, so it is possible to capture all relevant details in a yet concise model. Due to its powerful modeling capability, SPNs have become a useful tool for researchers in computer science.

Despite of its many advantages, one major challenge in using SPNs for performance evaluation is that the models developed in this way tend to result in Markov processes which have a large number of states. This phenomenon is known as the "state space explosion" problem, stemming from the fact that even very "innocent" nets, with a small number of places and transitions, can lead to very large state spaces. Moreover, the state space of any Markovian based model generally grows exponentially with the dimensions of the modeled system, making it impossible to obtain any solution for large scale systems. In order to deal with this problem, a set of powerful techniques have been developed in SPNs, such as model decomposition and model aggregation, which will be discussed in detail in the next chapter.

SPNs have received relatively little attention in the research of wireless networks. This is mostly due to the fact that SPNs are unknown to most wireless researchers. The purpose of this book is to introduce the basic principles of SPNs and show its applications to the performance evaluation of wireless networks with several examples.

1.3   Related Works on SPNs for Wireless Networks

In recent years, SPNs have been used occasionally to model wireless communications systems, but a widespread use is not observed. In this section, we provide a brief overview of the existing state-of-art research.

1.3.1   Ad Hoc Networks

In [12], the authors present an approach for the modeling and analysis of large-scale ad hoc networks using SPNs. In order to deal with the state space explosion problem, an approximate model is proposed based on the idea of SPN decomposition, which exploits the large amount of nodes and essentially describes the behavior of one node under a workload that is generated by the whole ad hoc network. It is shown that a close match exists between the obtained results by SPN approach and those from a simulation model in ns2. Moreover, the proposed scheme costs negligible computational effort compared with that of a simulation method.

In [13], the authors present a performance study of the Distributed Coordination Function (DCF), which is the fundamental contention-based access mechanism of 802.11 wireless LANs. The proposed SPN model can capture all relevant system aspects in a concise way due to ability of SPN formalisms in structuring complex models into modules. Numerical results obtained from the SPN model allow to quantify the influence of many mandatory features of the standard on performance, especially the backoff procedure, extended interframe spaces, and the timing synchronization function.

1.3.2   Cellular Networks

In [14], the authors address the scheduling tradeoff between average cell spectral efficiency, cell edge performance and fairness in cellular networks. SPNs are used to model different schedulers, so that their performance can be obtained by Markov chain steady state analysis without simulation. In an extended work [15], the authors treat the scheduling tradeoff analysis with a mix of Realtime (RT) and Non-Realtime (NRT) traffic. The system is abstractly modeled as a SPN which incorporates a parameter that models the tuning parameter of a real scheduler. The numerical results show that a tradeoff is possible only for NRT traffic, but with increasing proportion of RT traffic, this flexibility shrinks down to zero.

In [16], the authors propose an efficient Call Admission Control (CAC) scheme for mobile networks that takes into account voice connections as well as synchronous and asynchronous data connections. Since Stochastic Well-Formed Petri Nets (SWNs) are a powerful tool for modeling complex systems with concurrency, synchronization and cooperation, they are used to model the system interaction, which consists of several mobile nodes, gateways, cells, and servers. In [17], the authors develop an executable top-down hierarchical Colored Petri Net (CPN) model for multi-traffic CAC in Orthogonal Frequency Division Multiple Access (OFDMA) system. Moreover, four CAC schemes are presented based on the CPN model taking into account call-level and packet-level QoS. The simulation results show that CPN offers significant advantages over Markov chain in modeling CAC strategies and evaluating their performance with less computational complexity in addition to its flexibility and adaptability to different scenarios.

In [18], the authors focus on modeling, performance evaluation and reliability of small cell wireless networks, taking into account the retrial phenomenon, finite number of customers served in a cell and channels breakdowns. It is shown that the Generalized Stochastic Petri Net (GSPN) model can be used to cope with the complexity of such finite-source retrial networks under different breakdowns disciplines, and to derive the related performance and reliability indices.

In [19], the authors present a Stochastic Reward Net (SRN) model for performance evaluation of bandwidth allocation in IEEE 802.16 network considering multiple traffic classes. The proposed model incorporates prioritization and preemption of traffic classes. Packet drop due to waiting time exceeding threshold is also considered. The performance of the system is evaluated in terms of mean delay and normalized throughput considering the on-off traffic model.

1.3.3   Multi-hop Wireless Networks

In [20], the authors use SPN formalisms to build different channel models, such as the Gilbert-Elliot (GE) channel and the FSMC. Wireless system models can be formulated using the SPN-based channel models, so the performance of single-hop system and multi-hop system models can be analyzed. In [21], a credit-based flow control protocol from the ATM domain is proposed to be used in multi-hop wireless networks. The authors study the performance of the flow control protocol on two-hop wireless networks by making use of the SPN-based channel model discussed in [20]. It is shown that SPN model for two-hop wireless networks can be easily extended to multiple hops and even towards mesh networks.

References

1. R. Agrawal, V. Subramanian (2002) Optimality of certain channel aware scheduling policies. Paper presented at the 40th Conference on Communication, Control and Computing, 2002
2. A. Stoyar (2005) On the asymptotic optimality of the gradient scheduling algorithm for multiuser throughput allocation. Operations Research 53:12–25
3. M. Andrews (2004) Instability of the proportional fair scheduling algorithm for HDR. IEEE Trans. Wireless Commun 3(5):1422–1426
4. M. Andrews et al (2004) Scheduling in a queueing system with asynchronously varying service rate. Probability in the Engineering and Informational Sciences 18:191–217
5. A. Eryilmaz, R. Srikant, J. R. Perkins (2005) Stable scheduling policies for fading wireless channels. IEEE/ACM Trans. Netw 13(2):411–424
6. M. Dianati, X. Shen, S. Naik (2007) Scheduling with base station diversity and fairness analysis for the downlink of CDMA cellular networks. Wireless Communications and Mobile Computing (Wiley) 7:569–579
7. M. Dianati, X. Shen, S. Naik (2007) Cooperative fair scheduling for the downlink of CDMA cellular networks. IEEE Trans. on Vehicular Technology 56(4):1749–1760
8. M. Andrews (2005) A survey of scheduling theory in wireless data networks. Paper presented at the IMA summer workshop on wireless communications, 2005
9. S. C. Borst (2003) User-level performance of channel-aware scheduling algorithms in wireless data networks. IEEE/ACM Trans. on Networking 13(3):636–647
10. T. Bonald, S. C. Borst, A. Proutiere (2004) How mobility impacts the flow-level performance of wireless data system. Paper presented at the 23rd AnnualJoint Conference of the IEEE Computer and Communications Societies, Hong Kong, 7–11 March 2004
11. R. Prakash, V. V. Veeravalli (2007) Centralized wireless data networks with user arrivals and depatures. IEEE Trans. on Inf. Theory 53(2):693–713
12. C. Zhang, M. Zhou (2003) A stochastic Petri net-approach to modeling and analysis of ad hoc network. Paper presented at the Proceedings. ITRE2003 International Conference on Information Technology: Research and Education, 11–13 Aug 2003
13. Armin Heindl, Reinhard German (2001) Performance modeling of IEEE 802.11 wireless LANs with stochastic Petri nets. Performance Evaluation 139–164
14. Rainer Schoenen, Akram Bin Sediq, Halim Yanikomeroglu (2011) Fairness Analysis in Cellular Networks using Stochastic Petri Nets. Paper presented at IEEE 22nd International Symposium on Personal Indoor and Mobile Radio Communications (PIMRC), Toronto, 11–14 Sept 2011
15. R. Schoenen, A. B. Sediq, H. Yanikomeroglu et al (2012) Spectral efficiency and fairness tradeoffs in cellular networks with realtime-nonrealtime traffic mix using stochastic Petri nets. Paper presented at the 2012 IEEE VTC-Fall, 3–6 Sept 2012
16. Lynda Mokdad, Mbaye Sene, Azzedine Boukerche (2011) Call Admission Control Performance Analysis in Mobile Networks Using Stochastic Well-Formed Petri Nets. IEEE Transactions On Parallel And Distributed Systems 22(8):1332–1341
17. Yao Yuanyuan, Lu Yanhui, Yang Shouyi (2012) Modeling Multi-traffic Admission Control In OFDMA System Using Colored Petri Net. Journal Of Electronics 29(6):509–514

18. Nawel Gharbi (2012) Modeling and Performance Evaluation of Small Cell Wireless Networks with Base Station Channels Breakdowns. Paper presented at the 8th International Conference on Wireless and Mobile Communications, 2012
19. Shanmugam Geetha, Raman Jayaparvathy (2010) Modeling and Analysis of Bandwidth Allocation in IEEE 802.16 MAC: A Stochastic Reward Net Approach. Int. J. Communications, Network and System Sciences 631–637
20. R. Schoenen, M. Salem, A. Sediq et al (2011) Multihop wireless channel models suitable for stochastic Petri nets and Markov state analysis. Paper presented at the IEEE 73rd Vehicular Technology Conference (VTC Spring), Budapest, 15–18 May 2011
21. Rainer Schoenen (2011) Credit-Based Flow Control for Multihop Wireless Networks and Stochastic Petri Nets Analysis. Paper presented at the 9th Communication Networks and Services Research Conference, 2–5 May 2011

Chapter 2
Stochastic Petri Nets

SPNs were introduced in 1980 as a formalism for the description of discrete event systems whose dynamic behavior can be represented by means of continuous-time homogeneous Markov chains [1]. Although SPN models are widely used for performance and reliability evaluation of many practical systems, the major problem in this approach, however, is that a large state space of the underlying Markov model needs to be generated, stored, and processed. In this chapter, we will introduce two powerful techniques in SPNs to deal with the above state space explosion problem: one is the model decomposition and iteration technique; the other is the model aggregation technique in SHLPNs.

2.1 SPN Basics

Before introducing SPNs, it is necessary to learn about what are PNs. PNs were developed originally by Carl Adam Petri in 1962. Since then, they have been extended and developed, and applied in a variety of areas, such as office automation, manufacturing, programming design, computer networks, communications, Internet, railway networks and biological systems.

A PN is a directed bipartite graph with two types of nodes called *places* and *transitions* which are represented by circles and rectangles (or bars), respectively. Arcs connecting places to transitions are referred to as *input arcs*; while the connections from transitions to places are called *output arcs*. A non-negative integer (the default value is one) may be associated with an arc, which is referred to as *multiplicity* or *weight*. Places correspond to state variables of the system, while transitions correspond to actions that induce changes of states. A place may contain tokens that are represented by dots in the PN. The state of the PN is defined by its marking, which is represented by a vector $\mathbf{M} = (l_1, l_2, \ldots, l_k)$, where $l_k = M(p_k)$

© The Author(s) 2015
L. Lei et al., *Stochastic Petri Nets for Wireless Networks*, SpringerBriefs
in Electrical and Computer Engineering, DOI 10.1007/978-3-319-16883-8_2

is the number of tokens in place p_k. Here, $M(\cdot)$ is a mapping function from a place to the number of tokens assigned to it.

A PN is formally defined by the following tuple

$$PN = (P, T; F, W, \boldsymbol{m_0}) \qquad (2.1)$$

where

- $P = (p_1, p_2, \cdots, p_P)$ is the set of places.
- $T = (t_1, t_2, \cdots, t_T)$ is the set of transitions.
- $F \subseteq (P \times T) \cup (T \times P)$ is the set of arcs which are between places and transitions (and between transitions and places).
- $W : F \rightarrow \mathbb{N}$ is a weight function.
- $\boldsymbol{m_0} = (m_{01}, m_{02}, \cdots, m_{0P})$ is the initial marking.

Figure 2.1 is a simple PN with all components.
This PN has two places P_1, P_2, and one transition T_1. P_1 has one token and P_2 has no token, that is $M(P_1) = 1$, $M(P_2) = 0$.

A transition must be enabled before it is fired. A transition is enabled when the number of tokens in each of its input places is at least equal to the arc weight going from the place to the transition. An enabled transition may fire at any time. When fired, the tokens in the input places are moved to output places, according to arc weights and place capacities. This results in a new marking or state of the PN. Figure 2.2 is an example of firing a transition.

Fig. 2.1 A simple Petri Net

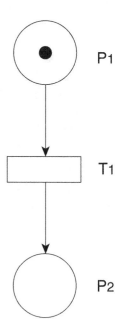

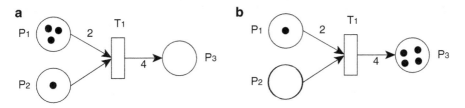

Fig. 2.2 Petri Nets transition process. (**a**) Petri Nets before T_1 fires, (**b**) Petri Nets after T_1 fires

SPNs are one kind of PNs in which an exponentially-distributed time delay is associated with each transition. An SPN is formally defined by the following six-tuple

$$SPN = (P, T; F, W, m_0, \lambda) \tag{2.2}$$

where $(P, T; F, W, m_0)$ have the usual meanings so that the underlying PN model constitutes the structural component of a SPN model. $\lambda = (\lambda_1, \lambda_2, \cdots, \lambda_T)$ is the set of firing rates associated with transitions, where λ_i is the mean firing rate of transition t_i. The reciprocal of firing rate $\tau_i = 1/\lambda_i$ denotes the mean firing delay or mean service time of transition t_i.

SPNs can be associated with stochastic process. Due to the memoryless property of the exponential distribution of firing delays, it is easy to find that SPN systems are isomorphic to Continuous-Time Markov Chains (CTMCs). Specifically, a state in the Markov process is associated with every marking in the SPNs. In addition, an event, or transition, in the Markov process is associated with each firing of a transition in the SPNs which causes the corresponding change of marking.

Although SPNs provide a clear and intuitive formalism for generating Markov processes, they do have the disadvantage that the models constructed in this way can soon become exceedingly large. GSPNs represent an extension of the SPN formalism, which are designed to address this problem. GSPNs divide the transitions into two classes: the exponentially-distributed timed transitions (represented by blank rectangles), which are used to model the random delays associated with the execution of activities, and immediate transitions (represented by bars), which are devoted to the representation of logical actions that do not consume time.

The immediate transitions have precedence over immediate transitions in firing. Consider the example in Fig. 2.3. A token in place P_1 starts the activity modeled by timed transition T_1. If a token arrives in P_2 before the firing of T_1, immediate transition T_2 becomes enabled and fires, thus disabling timed transition T_1.

The presence of immediate transitions brings about a difference among markings. Markings in which no immediate transitions are enabled are called tangible, however markings enabling at least one immediate transition are referred to as vanishing. The GSPN system spends a positive amount of time in tangible markings, and a null time in vanishing markings. Since the vanishing markings can be

Fig. 2.3 Petri Nets with an immediate transition

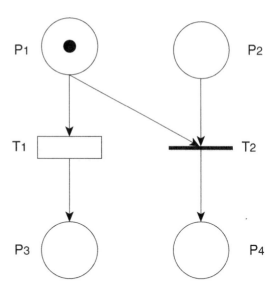

eliminated from the reachability graph before the Markov process is generated, the GSPN models have smaller number of markings compared with the SPN models.

Due to the space limitation of this book, we cannot provide a comprehensive introduction of the SPNs and GSPNs. Interested readers may refer to [1, 2] for more detailed information. Here, we only introduce two specific forms of SPNs that will be used in the following chapters. The first one is the Deterministic and Stochastic Petri Nets (DSPNs), which further extend GSPNs in that they allow timed transitions to have an exponentially-distributed time delay or an deterministic timed delay (represented by filled rectangles). The second one is the Discrete Time Stochastic Petri Nets (DTSPNs) [3], which are extended from the PNs by associating a geometrically distributed delay with each transition, so that they can be mapped to a Discrete-Time Markov Chain (DTMC) instead of CTMC as in regular SPNs. A nonzero conditional probability p is associated with each transition. The probability $p < 1$ is defined as the probability that the enabled transition fires at the next time step, given (conditioned on) the fact that no other transition fires. Since multiple firings may occur at any time step in a DTSPN, its analysis requires an additional step to decondition the probabilities before the normal Markovian analysis is attempted.

2.2   Model Decomposition and Iteration Technique

To make performance evaluation of SPNs attractive for many real life applications, we have to deal with models in the range of billions of states. Model decomposition uses the "divide and conquer" principle in various ways to overcome complexity.

It breaks up a model according to its network structure and divides the model into multiple submodels. Since the solution complexity is roughly exponential in the model size, it is cheaper to solve several small models than a single large one. Even if each smaller model must be solved several times in an iteration, the total effort can be orders of magnitude less [4]. The price is an approximation in performance measures.

Several model decomposition techniques have been proposed. For example in [5], the time scale decomposition technique was proposed by Ammar and Islam in 1989. It is suitable for GSPN models which are described by system of different magnitude operation time. When it comes to the reliability, efficiency and dependency of system, there may be orders of magnitude differences in firing rate of transitions. For instance, there is a big difference between firing rate of normal operation and that of failure operation: since a failure operation is rare, its firing rate is much smaller. In this section, we will focus on another important model decomposition technique developed by Ciardo and Trivedi in 1993 [6], which decomposes a GSPN into a set of "nearly independent" subnets and solves each individual subnet separately. Because there exist some dependencies among the subnets, after solving each subnet, certain quantities need to be exported to other subnets, and this is conducted iteratively.

When decomposing a Markov chain to obtain an approximate steady-state solution, the quality of the approximation is related to the degree of coupling among the blocks into which the Markov matrix is decomposed. Better approximations are obtained when the transition rate matrix is near-decomposable, i.e, its off-diagonal blocks are close to zero. In contrast to the above Markov chain level decomposition, it is more desirable to decompose a GSPN model directly (net level decomposition) without generating the underlying Markov chain, whose size is the main limitation to study the GSPN. In [6], the concept of a near-independent Markov chain is proposed to characterize the level of approximation when applying the net level decomposition. For example, consider the GSPN $\mathbf{C}$, obtained by composing two independent GSPNs $\mathbf{A}$ and $\mathbf{B}$. Construct $R^{\mathbf{C}}$ and $S^{\mathbf{C}}$, the transition rate matrix of the underlying CTMC and the tangible reachability set of the GSPN $\mathbf{C}$, and compare them to $R^{\mathbf{A}}$, $S^{\mathbf{A}}$, $R^{\mathbf{B}}$ and $S^{\mathbf{B}}$, the transition rate matrices and the tangible reachability sets of $\mathbf{A}$ and $\mathbf{B}$. If $|S^{\mathbf{A}}| = m$ and $|S^{\mathbf{B}}| = n$, then $R^{\mathbf{A}}$ is $m \times m$ and $R^{\mathbf{B}}$ is $n \times n$. $S^{\mathbf{C}}$ is the cross-product $S^{\mathbf{A}} \times S^{\mathbf{B}}$. $R^{\mathbf{C}}$ can be expressed as the Kronecker sum [7]

$$R^{\mathbf{C}} = R^{\mathbf{A}} \oplus R^{\mathbf{B}} = R^{\mathbf{A}} \otimes I_n + I_m \otimes R^{\mathbf{B}},$$

where I_i is the $i \times i$ identity matrix. We recall that the Kronecker product "$\otimes$" of two matrices E, $e_r \times e_c$, and F, $f_r \times f_c$, is the $e_r f_r \times e_c f_c$ matrix

$$E \otimes F = \begin{bmatrix} E_{1,1}F & \cdots & E_{1,e_c}F \\ \cdots & \cdots & \cdots \\ E_{e_r,1}F & \cdots & E_{e_r,e_c}F \end{bmatrix}$$

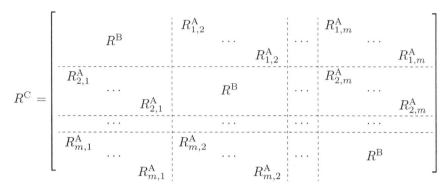

Fig. 2.4 The structure of R^{C}

R^{C} consists of m_2 blocks of size $n \times n$ (Fig. 2.4). Each of the m diagonal blocks is equal to R^{B}, while the off-diagonal block in position (i, j) has zero entries except on the diagonal where all the elements are equal to $R^{\mathrm{C}}_{i,j}$

The Kronecker sum and product are noncommutative operators, but $R^{\mathrm{A}} \oplus R^{\mathrm{B}}$ and $R^{\mathrm{B}} \oplus R^{\mathrm{A}}$ differ only in the ordering of the indices.

A transition rate matrix is near-independent if it can be partitioned so that the diagonal contains two or more occurrences of (approximately) the same block, while the off-diagonal blocks must have (approximately) the structure of Fig. 2.2, but they are not at all required to have small entries as the near-decomposable matrix. Three basic structures of near-independent GSPN models are proposed in [6].

- **Rate relation**—For an SPN model consisting of multiple submodels, at least one rate in one submodel is a function of the marking of another submodel. This will lead to one of the positive entries of the transition rate matrix changing to a different positive value.
- **Synchronization relation**—There is at least one synchronous transition between different submodels, which results in one of the off-diagonal zero entries in an off-diagonal block changing to a positive value.
- **External relation**—There exists at least one forbidden arc between different submodels, which results in one of the diagonal positive entries in an off-diagonal block changing to zero.

We generally need the following steps to exploit near-independence at the GSPN level.

1. **Decomposition**—Given a GSPN A, generate GSPNs $A_1, \cdots A_k$. Often, A_i is not a subnet of A, but it shares common subnets with it. These GSPNs are parametrized: the firing rates of some of their transitions might be expressed as a function of (real) parameters to be specified.
2. **Import graph**—For each SPN A_i, fix the value of its parameters using *imports* from A_j, $1 \le j \le k$. Also, specify the quantities exported from A_i after its

solution. If A_i imports a quantity from A_j, we write $A_j \succ A_i$. With a slight abuse of notation, we also denote the transitive closure of the import relation with the symbol $\succ$. The graph describing the import relation may be cyclic (the case $A_i \succ A_i$, a cycle of length one, may occur).

3. **Iteration**—If the import graph is acyclic, it implicitly defines a (partial) order for the solution of the GSPNs A_i. If $A_i \succ A_j$, Ai must be studied before A_j. If neither $A_i \succ A_j$ nor $Aj \succ Ai$, then A_i and A_j can be studied in any order. If A_i and A_j belong to a cycle, one of them must be chosen to be studied first, but its imports are not available and initial guesses must be provided for them. After each A_i has been studied once, more iterations can be performed, each time using the most recent value for the imports. Convergence is reached when all the imports remain (almost) constant between successive iterations.

Note that the iteration in Step 3 is necessary because the model cannot be decomposed very "cleanly", i.e., there are interactions between submodels that cannot be ordered. In such cases fixed point iteration is used to determine those model parameters that are not available directly as input or by solving other models. In this technique, the relations between model parameters and model outputs result in an equation of the type

$$\mathbf{x} = f(\mathbf{x}), \tag{2.3}$$

where $\mathbf{x} := (x_1, \ldots, x_n)$ is the vector of iteration variables. This is the fixed point equation corresponding to the iterative model, and the vector $\mathbf{x}$ that satisfies this equation is called a fixed point of this equation. The simplest way of finding this fixed point is by successive substitution. In this method, starting with an initial guess $\mathbf{x}_0$ we iterate in the following way:

$$\mathbf{x}_n = f(\mathbf{x}_{n-1}). \tag{2.4}$$

This iteration is terminated when the difference between two successive iterates is below a certain tolerance level. Note that this iteration may not always converge. If it converges, it may not always converge to the same value. Therefore, before we use the iterative method, we must be sure that a solution to the above equation exists. In [8], the theoretical problems that arise while using iteration with SPNs are studied. Specifically, it is proved that if the iteration variables are expected reward rates and the underlying CTMC has exactly one closed communicating class, a fixed point will exist.

2.3 Stochastic High-Level Petri Nets

The model decomposition techniques have been widely used in SPNs and can work quite well in practice. They drastically reduce the solution time with respect to that of the overall exact model (which might not even be feasible to solve).

However, the model decomposition techniques can only be applied to the models with specific structures, e.g., near-independency. For those SPN models without these desirable structures, another powerful technique that may be considered is the model aggregation, which corresponds to the state grouping in the Markov domain. In this section, we will introduce one such model aggregation technique— the compound marking technique in SHLPNs.

SHLPNs are extensions of High-Level Petri Nets (HLPNs) in which each transition has an exponentially distributed firing time associated with it [9]. HLPNs, on the other hand, are extensions of regular PNs that lead to simpler models with a more readable graph. Specifically, markings in regular PNs are indiscriminate, while markings in HLPNs may have properties or be distinguished by different colors. Different types of HLPNs have been proposed, for example, predicate transition nets, colored Petri nets, relation nets, and all of them are conceptually similar. Moreover, the model of a system constructed using one type of HLPN can be informally translated into any other type of HLPN.

A HLPN consists of the following elements.

(a) A directed graph (P, T, A) where

- P is the set of places
- T is the set of transitions
- A is the set of arcs, $A \subset (P \times T) \cup (T \times P)$.

(b) A structure set Σ consisting of some types of individual tokens (u_i) together with some operations (op_i) and relations (r_i), i.e., $\Sigma = (u_1, \ldots, u_n; op_1, \ldots, op_m; r_1, \ldots, r_k)$.

(c) A labeling of arcs with a formal sum of n attributes of token variables (including the zero-attributes indicating a noargument token).

(d) An inscription on some transitions being a logical formula constructed from the operation and relations of the structure Σ and variables occurring at the surrounding arcs.

(e) A marking of the places of P with n attributes of individual tokens.

(f) A natural number K which assigns to the places an upper bound for the number of copies of the same token.

(g) Firing rule: Each element of T represents a class of possible changes of markings. Such a change, also called transition firing, consists of removing tokens from a subset of places and adding them to other subsets according to the expressions labeling the arcs. A transition is enabled whenever, given an assignment of individual tokens to the variables which satisfies the predicate associated with the transition, all input places carry enough copies of proper tokens, and the capacity K of all output places will not be exceeded by adding the respective copies of tokens. The state space of the system consists of the set of all markings connected to the initial marking through such occurrences of firing.

The definition of SHLPNs is the combination of HLPNs and SPNs. The basic idea of SHLPNs is not only to keep the properties of HLPNs in model description and solution, but also have the state space and Markov Chain (MC) isomorphism as in SPNs by inducing exponentially distributed firing times to the transition set of HLPNs.

The state space size of SHLPN models can be further simplified due to the introduction of the compound marking technique, where subsets of equivalent states in the SPN models of homogeneous systems can be grouped together into a single compound state in the SHLPN model. The compound marking concept is based on the fact that a number of entities processed by the system exhibit an identical behavior and they have a single subnet in the SHLPN model. The only distinction between such entities is the identity attribute of the token carried by the entity. If, in addition, the system consists of identical processing elements distinguished only by the identity attribute of the corresponding tokens, it is possible to lump together a number of markings in order to obtain a more compact SHLPN model of the system. Clearly, the model can be used to determine the global system performance in case of homogeneous systems when individual elements are indistinguishable. Since there is an isomorphism between SHLPN and Markov chains, any compound markings of an SHLPN correspond to grouping, or lumping of states in the Markov domain.

Definition 2.1. A compound marking of a SHLPN is the result of partitioning an individual SPN marking into a number of disjoint sets such that:

- The individual markings in a given compound marking have the same distribution of tokens in places, except for the identity attribute of tokens of the same type.
- All individual markings in the same compound marking have the same transition rates to all other compound markings.

In this case, an equivalence relation exists among the SHLPN model with compound markings and the original SPN model with individual markings. Both provide the same information about the system being modeled but the SHLPN model with the compound marking is a scaled down version of the original SPN model with a lower number of states. Therefore, the main advantage of modeling homogeneous systems using SHLPN is that the resulting models are simpler, more intuitive, and have a smaller number of states.

We denote by p_{ij} the probability of a transition from the compound marking i to the compound marking j and by $p_{i_n j_k}$ as the probability of a transition from the individual marking i_n to the individual marking j_k, where $i_n \in i$ and $j_k \in j$. The relation between the transition probability of compound markings and the transition probability of individual markings is

$$p_{ij} = \sum_k p_{i_n j_k *}.$$

$$(2.5)$$

The relation between the transition rate of compound markings and the transition rate of individual markings is

$$q_j(t) = \frac{d\left(\sum_i p_{ji}\right)}{dt} = \frac{\sum_i d\left(\sum_k p_{j_n i_k}\right)}{dt}, \tag{2.6}$$

$$q_{ij}(t) = \frac{dp_{ij}}{dt} = \frac{\sum_k d\left(p_{i_n j_k}\right)}{dt}. \tag{2.7}$$

If the system is ergodic, then the sojourn time in each compound marking is an exponentially distributed random variable with average

$$\left[\sum_{i \in H} (q_{jk})_i\right]^{-1}, \tag{2.8}$$

where H is the set of transitions that are enabled by the compound marking and q_{jk} is the transition rate associated with the transition i firing on the current compound marking j.

References

1. G. Balbo (2001) Introduction to Stochastic Petri Nets. Lecture Notes in Computer Science 2090:84–155
2. P. J. Haas (2002) Stochastic Petri Nets: Modelling, Stability, Simulation. Springer, Heidelberg
3. M. K. Molloy (1985) Discrete Time Stochastic Petri Nets. IEEE Trans. Software Engineering 11:417–423
4. Y. Li, C. M. Woodside (1995) Complete Decomposition of Stochastic Petri Nets Representing Generalized Service Networks. IEEE Trans. Comput 44:1031–1046
5. H. H. Ammar, S. M. R. Islam (1989) Timed scale decomposition of a class of generalized Stochastic Petri Net models. IEEE Transactions on Software Engineering 15:809–820
6. G. Ciardo, K. S. Trivedi (1993) A decomposition approach for stochastic reward net models. Performance Evaluation 18:37–59
7. V. Amoia, G. De Micheli, M. Santomauro (1981) Computer-oriented formulation of transition-rate matrices via Kronecker algebra. IEEE Trans 30(2):123–132
8. V. Mainkar, K. S. Trivedi (1995) Fixed point iteration using stochastic reward nets. Paper presented at the sixth International Workshop on Petri Nets and Performance Models, 1995
9. C. Lin, D. C. Marinescu (1988) Stochastic High-Level Petri Nets and Applications. IEEE Trans. Comput 37:815–825

Chapter 3
Performance Analysis of Opportunistic Schedulers Using SPNs

In Chap. 2, we have introduced the model decomposition and iteration technique in SPNs to deal with the state space explosion problem. In this chapter, we adopt this technique to study the performance of wireless opportunistic schedulers in multiuser systems under a dynamic data arrival setting. We first develop a framework based on Markov queueing model and then analyze it by applying the decomposition and iteration technique. Since the state space size in our analytical model is small, the proposed framework shows an improved efficiency in computational complexity. Based on the established analytical model, performance of both opportunistic and non-opportunistic schedulers are studied and compared in terms of average queue length, mean throughput, average delay and dropping probability. Analytical results demonstrate that the multiuser diversity effect as observed in the infinite backlog scenario is only valid in the heavy traffic regime. The performance of the Channel-Aware (CA) opportunistic schedulers is worse than that of the non-opportunistic round robin scheduler in the light traffic regime, and becomes worse especially with the increase of the number of users. Simulations are also performed to verify the accuracy of the analytical results.

3.1 Packet Level Performance Analysis of Opportunistic Schedulers

In wireless systems, channel conditions are inherently time-varying due to the existence of fading and shadowing effects. Moreover, since different wireless users may experience independent channel variations, the event that there exist users with strong channel gains at any time instant occurs with high probability, which is referred to as *multiuser diversity*. In order to exploit such channel fluctuation and multiuser diversity for throughput improvement in wireless systems, Opportunistic Scheduling (OS) has been appeared. Here, the "opportunistic" means

© The Author(s) 2015
L. Lei et al., *Stochastic Petri Nets for Wireless Networks*, SpringerBriefs
in Electrical and Computer Engineering, DOI 10.1007/978-3-319-16883-8_3

the mechanism can take advantages of the favorable channel conditions in resource allocation. So far, the concept of OS has been widely applied in the 3rd-Generation (3G) wireless systems such as Code Division Multiple Access (CDMA) 2000 1xEV-DO [1] and 4G wireless systems such as 3rd Generation Partnership Project (3GPP) Long Term Evolution (LTE). While all OS algorithms take into account the channel state information, some of them may also consider the queueing status of users. In this chapter, we use "channel/queue-aware" and "channel-aware" to indicate respectively whether queue state information is considered or not in OS algorithms.

We focus on packet-level performance analysis of OS algorithms considering dynamic packet arrivals. Several research work in this aspect focused on network stability, i.e., the queue occupancy can be bounded whenever feasible [2, 3], with both channel-aware and channel/queue-aware OS algorithms. The typical observations are that most channel-aware OS algorithms, e.g., the Proportional Fair (PF) algorithm, are unstable [2] and a quadratic Lyapunov function argument can be used to prove the stability for channel/queue-aware algorithms [3]. Moreover, several research work also studied the statistical worst case performance of OS algorithms using effective bandwidth and its related concepts [4–6]. In [5], a formula is provided to approximate the tail distribution of packet delay for the greedy channel-aware and round-robin algorithms under the FSMC model. In [6], the maximum throughput for the channel-aware and the channel/queue-aware algorithms are estimated under the constraints that the tail distribution of the queue length cannot exceed a certain threshold, where the wireless channel condition or variation is assumed to be a memoryless on-off process.

While all these works on the network stability and statistical worst case performance provide important insights into the queueing behavior of the OS algorithms, the average performance, such as average delay and average throughput, are also essential in network design. However, the difficulties lie in deriving the steady-state distribution of the queue states. This situation becomes even harder in wireless systems due to the time-varying channel conditions. In [7], a two-dimensional Markov model for computing the steady-state distribution in single user systems is proposed, where two dimensions represent channel and queue states, respectively. However, such analytical model cannot be directly extended to multiuser systems since the state space of the Markov model grows exponentially as the number of users increases. Thus, designing a practically implementable analysis model is critical for analyzing the performance of OS algorithms in multiuser wireless systems.

In this chapter, a new analytical framework is proposed for multiuser systems, which can be used for studying the performance of different wireless schedulers in terms of average queue length, mean throughput, average delay and dropping probability. The wireless schedulers under consideration include not only opportunistic schedulers using channel-aware and channel/queue-aware algorithms, but also non-opportunistic ones using round-robin and queue-aware algorithms. Specifically, by characterizing the service process from the FSMC as a Markov Modulated Deterministic Process (MMDP), the wireless downlink for the multiuser system is first modeled as an M/MMDP/1/K queueing system. Then, a deterministic & DSPNs model is constructed, where different scheduling algorithms can be expressed by model parameters, such as the enabling predicates and random switches. By applying the model decomposition technique in SPNs, the multiuser system is decomposed into multiple single user subsystems with inter-correlated service rates. To facilitate the analysis for each subsystem separately, some approximation methods including the replacement of the instantaneous service rates by steady-state average ones and a fixed-point iteration method, are introduced. The proposed analytical framework significantly reduces the state space of the Markovian system model in analysis and shows good performance in scalability. Numerical results show that (1) the channel-aware algorithm performs better than the round robin algorithm only in heavy traffic regime; (2) the scheduling gain of the channel-aware algorithm increases with the number of users only when the traffic load per user is heavy; and (3) the channel/queue-aware algorithm outperforms the channel-aware algorithm in light traffic regime and converges to the channel-aware algorithm in heavy traffic regime.

Notice that the selection of stochastic Petri nets approach results from the facts that (1) it provides an intuitive and efficient way in describing the multiuser system, especially facilitating the inclusion of different scheduling strategies; and (2) there exist a set of well-developed techniques for stochastic Petri nets, which can decompose the original complex model into simple subsystems and provide iteration methods for performance approximation.

3.2 The DSPN Model Formulation

In this section, a general framework is introduced for modeling and analyzing a multiuser system using DSPNs.

3.2.1 The M/MMDP/1/K Queuing System

Consider the downlink of a cellular wireless network, where a Base Station (BS) transmits data to $N(N \geq 1)$ mobile users. The BS maintains a separate data buffer for each mobile user and each buffer has a finite capacity of $K < \infty$ bits. For each data buffer n, the data arrives according to Poisson distribution with average rate λ_n bits/s. The transmission in the time is slot-by-slot based and each slot has an equal length ΔT. In each time slot the BS can transmit data to one user only. It is assumed that all channel conditions are available at the BS so that the Adaptive Modulation and Coding (AMC) algorithms can be applied. The wireless channel for each mobile user is modeled as an independent FSMC with total L states. Each state of FSMC corresponds to one non-overlapping consecutive Signal to Noise Ratio (SNR) region and a fixed transmission rate determined by the AMC algorithm. During each time slot, the channel stays at the same state.

The described system model can be formulated by an M/MMDP/1/K queueing system as follows. The defined fluid queueing system consists of a finite number, N, of input flows indexed by $n = 1, 2, \ldots, N$, and one server. For any n, there is a finite, irreducible, and continuous-time Markov chain $H_n(\tau)$ with total L server states, which corresponds to the L channel states of the FSMC model. The transition rates of the Markov chain depends on its channel fading speed and is not necessarily identical for all input flows, but the transitions of different Markov chains are independent. Associated with the l-th ($l \in 1, \ldots, L$) state of $H_n(\tau)$ is a fixed service rate $R_{n,l}$ bits/s, which is a non-negative integer. If at time τ the server is allocated to flow n with $H_n(\tau)$ in the l-th state, the queue n is served at a deterministic rate $R_{n,l}$, i.e., the user is served according to an L-state MMDP.

3.2.2 The DSPN Model

The M/MMDP/1/K queueing system designed for the multiuser wireless downlink can be equivalently modeled as a DSPN by following the similar procedure as shown in [8]. The modeled DSPN consists of a SPN for representing service processes and a DSPN for representing resource sharing. The SPN, as shown in Fig. 3.1a, is further composed of N subnets and each subnet n corresponds to the L-state Markov modulated service process of user n. Each subnet is described by places $(\{H_{nl}\}_{l=1}^{L})$ and transitions $(\{tu_{nl}\}_{l=1}^{L-1}$ and $\{td_{nl}\}_{l=1}^{L-1})$. The DSPN, as shown in Fig. 3.1b, models the resource sharing relationship of the multiuser system and can be characterized by places $(\{Q_n\}_{n=1}^{N}, \{w_n\}_{n=1}^{N}$ and $r)$ and transitions $(\{c_n\}_{n=1}^{N}, \{d_n\}_{n=1}^{N}$ and $\{s_n\}_{n=1}^{N})$. The meanings of all the places and transitions are described as follows.

- Q_n: a place for the queue state of user n.
- H_{nl}: a place for the l-th server state of user n.
- c_n: an exponentially-distributed timed transition denoting new bit arrivals from user n, with firing rate λ_n. When it fires, one bit of data arrives at the queue place Q_n.
- d_n: an immediate transition denoting the execution of scheduling strategy. Different scheduling strategies are expressed using different enabling predicates and random switches. The details will be discussed in the next subsection.
- s_n: a deterministic timed transition for service process. When it fires, one bit of data is transmitted from the queue place Q_n. Its firing rate μ_n depends on the marking of the places $\{H_{nl}\}_{l=1}^{L}$, i.e., if $M(H_{nl}) = 1$, then

$$\mu_n = R_{n,l}, \ l = 1, \ldots, L,$$

where $M(\cdot)$ is a mapping function from a place to the number of tokens assigned to it, and $M(H_n l)$ is either 1 or 0, which represents whether user n is in its l-th server state or not.

Fig. 3.1 The DSPN model
for multiuser wireless
downlink. (**a**) SPN for service
processes, (**b**) DSPN for
resource sharing

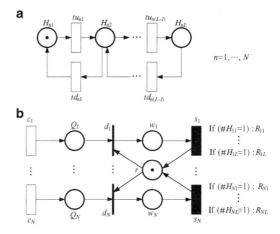

- tu_{nl}, td_{nl}: exponentially-distributed timed transitions for the server state transitions of user n. The firing rates of tu_{nl} and td_{nl} equal $p^n_{l,l+1}/\triangle T$ and $p^n_{l+1,l}/\triangle T$, which are determined by (3.35) and (3.36) in Appendix 2, respectively. When tu_{nl} (td_{nl}) fires, the server state transits from $l(l+1)$ to $l+1(l)$.

3.2.3 Scheduling Strategies

The performance of the multiuser system depends on the scheduling strategies applied. In the defined DSPN model, different scheduling strategies can be described by different enabling predicates and random switches of the immediate transition d_n. The enabling predicate specifies the condition under which user n is an eligible candidate for data transmission, while the random switch indicates the probability that user n will be selected for service.

1) *Round-Robin (RR) algorithm*: In round-robin algorithm, the scheduler polls queues for service in a cyclic order independent of the wireless channel conditions. Therefore, for the RR algorithm, the enabling predicate y_n of d_n is

$$y_n : (M(Q_n) > 0). \tag{3.1}$$

and the random switch $g_n(M)$ of d_n is

$$g_n(M) = \begin{cases} 1/\|RR(M)\|, & \text{if } n \in RR(M), \\ 0, & \text{otherwise,} \end{cases} \tag{3.2}$$

where $\mathbf{M}$ is a vector representing the number of tokens in each place of the DSPN model, which include $\{Q_n, \{H_{nl}\}^L_{l=1}\}^N_{n=1}$ and $RR(\mathbf{M}) = \{i \mid M(Q_i) > 0\}$.

2) *CA algorithms*: The channel-aware algorithms aim at improving the scheduling performance by incorporating channel state information. Two typical CA algorithms are greedy algorithm and PF algorithm.

   Greedy algorithm is also referred to as the Max-SNR algorithm. The algorithm always picks the user with the best SNR for transmission, or equivalently, the best transmission data rate is guaranteed at every scheduling instant. For the greedy algorithm, the enabling predicates and random switches can be found as

   $$y_n : (M(Q_n) > 0) \wedge (\forall i \neq n, \mu_i \leq n) \vee (\forall i \neq n, M(Q_i) = 0). \qquad (3.3)$$

   $$g_n(M) = \begin{cases} 1/\|CA(M)\|, & \text{if } n \in CA(M), \\ 0, & \text{otherwise,} \end{cases} \qquad (3.4)$$

   where $CA(M) = \{i \mid \mu_i = \max(\mu_1, \ldots, \mu_N), M(Q_i) > 0\}$, and the operators $\wedge$ and $\vee$ represent "logical and" and "logical or", respectively.

   PF algorithm, on the other hand, picks the user in each time slot among all backlogged users in the system which has the best transmission data rate normalized by the average throughput it has already received so far. Obviously, if all users experience statistically identical channels, there is no difference between the greedy algorithm and the PF algorithm in the long run, i.e., Eqs. (3.3) and (3.4) can also be used to describe the behavior of the PF algorithm. In fact, the enabling predication and random switches defined in (3.35) and (3.36) can be easily extended to describe the PF algorithm in more general scenarios by replacing the actual transmission rate μ_i, $i = 1, \ldots N$, with the normalized transmission rate $\mu_i/\bar{\mu}_i$, $i = 1, \ldots N$, where $\bar{\mu}_i$, $i = 1, \ldots N$, is the average transmission rate of user i.

3) *Queue-Aware (QA) algorithms*: The queue-aware algorithms are actually not "opportunistic" in the sense that they do not consider the channel state information in scheduling. Although they are more commonly used in wireline networks, in this chapter, for the purpose of comparison with the channel/queue-aware algorithms, a simple QA algorithm, which always selects the user with the largest queue length, is studied. For the QA algorithm, we have

   $$y_n : (M(Q_n) > 0) \wedge (\forall i \neq n, M(Q_i) \leq M(Q_n)). \qquad (3.5)$$

   $$g_n(M) = \begin{cases} 1/\|QA(M)\|, & \text{if } n \in QA(M), \\ 0, & \text{otherwise,} \end{cases} \qquad (3.6)$$

   where $QA(M) = \{i \mid M(Q_i) = \max(M(Q_1), \ldots, M(Q_N)), M(Q_i) > 0\}$.

4) *Channel/Queue-Aware (CQA) algorithms*: The systems with CA algorithms are ordinarily not stable, since the CA algorithms do not take into account the queue length information, and therefore do not know how to react when one queue starts getting too large. On the contrary, the CQA algorithms are a class of schedulers, which consider the information about both channel and queue occupancy. A

simple CQA algorithm is called the Max-Weight algorithm, where the user with maximum product of queue length and transmission data rate is served. For Max-Weight algorithm, we have

$$y_n : (M(Q_n) > 0) \wedge (\forall i \neq n, M(Q_i)\mu_i \leq M(Q_n)\mu_n). \tag{3.7}$$

$$g_n(M) = \begin{cases} 1/\|\text{CQA}(\mathbf{M})\|, & \text{if } n \in \text{CQA}(\mathbf{M}), \\ 0, & \text{otherwise}, \end{cases} \tag{3.8}$$

where $\text{CQA}(\mathbf{M}) = \{i \mid \mu_i M(Q_i) = \max(\mu_i M(Q_1), \ldots, \mu_N M(Q_N)), M(Q_i) > 0\}$.

3.3   Model Solution and Performance Analysis

In this section, analysis methods are introduced to find the solutions of the system derived in Sect. 3.2.

3.3.1   Single User System

We first consider the simplest case where there is only one mobile user, e.g., user n, in the system. Since previously defined M/MMDP/1/K queue underlying the DSPN model does not hold Markovian property, the direct analysis for such a single user system is still difficult. In order to simplify the analysis, we introduce the following embedded Markov chain [9]. Let $\mathcal{H}_n(\tau)$ be the Markovian server state process and $\mathcal{Q}_n(\tau)$ be the length of the queue occupancy at any time instant τ. Since the channel state keeps unchanged during each time slot, we can discretize the random processes at every ΔT interval and define $\mathcal{H}_{n,t} := \mathcal{H}_n(t \times \Delta T)$ and $\mathcal{Q}_{n,t} := \mathcal{Q}_n(t \times \Delta T)$. After discretization, the server and queue states are assumed to change only at the sample instant. Obviously, the two-dimensional embedded Markov chain $\{(\mathcal{H}_{n,t}, \mathcal{Q}_{n,t}), t = 0, 1, \ldots\}$ can accurately represent the system behavior. A similar discrete-time Markov chain has also been introduced in [7] for a single user wireless LAN (WLAN) system.

Let $p^n_{(l,k),(m,h)}$ be the transition probability from state (l, k) to state (m, h) of the embedded Markov chain. Then,

$$p^n_{(l,k),(m,h)} = \mathbb{P}\{\mathcal{Q}_{n,t+1} = h | \mathcal{H}_{n,t} = l, \mathcal{Q}_{n,t} = k\} p^n_{l,m} = v^{n,l}_{k,h} p^n_{l,m}, \tag{3.9}$$

where $v^{n,l}_{k,h} = \mathbb{P}\{\mathcal{Q}_{n,t+1} = h | \mathcal{H}_{n,t} = l, \mathcal{Q}_{n,t} = k\}$ and $p^n_{l,m}$ denotes the transition probability of the FSMC from state l to m. The determination of $p^n_{l,m}$ under Rayleigh fading channel is given in Appendix 1.

Let $A_{n,t}$ denote the number of bits arrived during the t-th time slot. Since the queueing process evolves following

$$\mathcal{Q}_{n,t+1} = \min[K, \max[0, \mathcal{Q}_{n,t} - R_{n,l}\Delta T] + A_{n,t}], \tag{3.10}$$

we have

$$v_{k,h}^{n,l} = \begin{cases} \mathbb{P}(A_{n,t} = h - k + R_{n,l}\Delta T) & k \geq R_{n,l}\Delta T, h \neq K, \\ \mathbb{P}(A_{n,t} = h) & k < R_{n,l}\Delta T, h \neq K, \\ \mathbb{P}(A_{n,t} \geq K - k + R_{n,l}\Delta T) & k \geq R_{n,l}\Delta T, h = K, \\ \mathbb{P}(A_{n,t} \geq K) & k < R_{n,l}\Delta T, h = K, \end{cases} \tag{3.11}$$

where $\mathbb{P}(A_{n,t} = u) = \frac{(\lambda_n \Delta T)^u}{u!} e^{-\lambda_n \Delta T}$ due to Poisson assumptions.

Define matrices $\boldsymbol{v}_{n,l} = [v_{k,h}^{n,l}]$ and $\mathbf{P}_n = [p_{(l,k),(m,h)}^n]$. We can partition $\mathbf{P}_n$ into blocks, each of which is a $(K+1) \times (K+1)$ matrix as shown in (3.12). Note that except for the main, upper and lower diagonals, all other blocks are zeros.

$$\mathbf{P}_n = \begin{pmatrix} p_{1,1}^n \boldsymbol{v}_{n,1} & p_{1,2}^n \boldsymbol{v}_{n,1} & & & & \\ p_{2,1}^n \boldsymbol{v}_{n,2} & p_{2,2}^n \boldsymbol{v}_{n,2} & p_{2,3}^n \boldsymbol{v}_{n,2} & & & \\ & p_{3,2}^n \boldsymbol{v}_{n,3} & p_{3,3}^n \boldsymbol{v}_{n,3} & p_{3,4}^n \boldsymbol{v}_{n,3} & & \\ & \ddots & \ddots & \ddots & & \\ & & p_{L-1,L-2}^n \boldsymbol{v}_{n,L-1} & p_{L-1,L-1}^n \boldsymbol{v}_{n,L-1} & p_{L-1,L}^n \boldsymbol{v}_{n,L-1} \\ & & & p_{L-1,L}^n \boldsymbol{v}_{n,L} & p_{L,L}^n \boldsymbol{v}_{n,L} \end{pmatrix}. \tag{3.12}$$

Define the steady-state probability $\pi_{l,h}^n \equiv \lim_{t \to \infty} \mathbb{P}\{\mathcal{H}_{n,t} = l, \mathcal{Q}_{n,t} = h\}$ and the vector $\boldsymbol{\pi}_n = (\pi_{1,0}^n, \pi_{1,1}^n, \dots, \pi_{1,K}^n, \dots, \pi_{L,0}^n, \pi_{L,1}^n, \dots, \pi_{L,K}^n)$. Then, the stationary distribution of the ergodic process $\{(\mathcal{H}_{n,t}, \mathcal{Q}_{n,t})\}$ can be uniquely determined from the balance equations

$$\boldsymbol{\pi}_n = \boldsymbol{\pi}_n \mathbf{P}_n, \quad \boldsymbol{\pi}_n \mathbf{e} = 1, \tag{3.13}$$

where $\mathbf{e}$ is the unity vector of dimension $L \times (K+1)$ and $\boldsymbol{\pi}_n$ can be derived as the normalized left eigenvector of $\mathbf{P}_n$ corresponding to eigenvalue 1. Given $\boldsymbol{\pi}_n$, the performance metrics such as the average queue length, the mean throughput, the average delay and the dropping probability can be derived.

- The average queue length equals

$$\overline{Q}_n = \sum_{k=0}^{K} \sum_{l=1}^{L} \pi_{l,k}^n k. \tag{3.14}$$

- The mean throughput can be expressed as

$$\overline{T}_n = \sum_{l=1}^{L} \sum_{k=1}^{K} T_{l,k}^n \pi_{l,k}^n, \tag{3.15}$$

where

$$T_{l,k}^n = \begin{cases} R_{n,l} & \text{if } k \geq R_{n,l}\Delta T, \\ \frac{k}{\Delta T} & \text{if } k < R_{n,l}\Delta T. \end{cases} \tag{3.16}$$

This is the sum of the product between the service rate in state l and the probability that the server is in state l given there are data in the system.

- The average delay then can be calculated as

$$\overline{D}_n = \overline{Q}_n / \overline{T}_n. \tag{3.17}$$

- Let $B_{l,k}^n$ be the random variable which represents the amount of dropped bits when $\mathcal{H}_{n,t} = l$ and $\mathcal{Q}_{n,t} = k$. Since $K + b = A_{n,t} + \max[0, k - R_{n,l}\Delta T]$, where b is the number of bits dropped during the t-th slot,

$$\mathbb{P}(B_{l,k}^n = b) = \mathbb{P}(A_{n,t} = K + b - \max[0, k - R_{n,l}\Delta T]). \tag{3.18}$$

Then, the dropping probability p_d^n can be estimated as

$$
\begin{aligned}
p_d^n &= \frac{\text{Average \# of bits dropped in a time slot}}{\text{Average \# of bits arrived in a time slot}} \\
&= \frac{\sum_{l=1}^{L} \sum_{k=0}^{K} \sum_{b=0}^{\infty} b \mathbb{P}(B_{l,k}^n = b) \pi_{l,k}^n}{\lambda_n \Delta T}.
\end{aligned}
\tag{3.19}
$$

3.3.2   Model Decomposition and Iteration

Although the analytical method in previous section for the single user system can be applied to the multiuser scenario by constructing an embedded Markov chain $\{\mathcal{H}_{1,t}, \mathcal{Q}_{1,t}, \ldots, \mathcal{H}_{N,t}, \mathcal{Q}_{N,t}\}$ with appropriately defined transition probabilities, the exponentially enlarged state space makes it unacceptable for a large user population. Since directly solving the DSPN suffers the high computational complexity, in this subsection, model decomposition and an iteration procedure are introduced to simplify the analysis.

Fig. 3.2 Decomposed DSPN model. (**a**) SPN for the service process of user n, (**b**) DSPN for the queuing process of user n

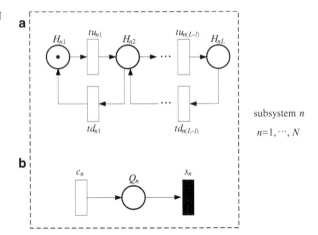

1) *Model decomposition*: According to Sect. 2.2, the original DSPN can be decomposed into a set of subnets, as shown in Fig. 3.2. The subnet Fig. 3.2a remains the same structure as that in Fig. 3.1a, while the DSPN in Fig. 3.1b is decomposed into N DSPN subnets in Fig. 3.2b. The n-th DSPN subnet consists of the exponentially-distributed timed transition c_n, the deterministic timed transition s_n, and the queue place Q_n. Note that the places r, w_n and the immediate transitions d_n in Fig. 3.1 are all deleted in the decomposed model for simple model description. The transition c_n and the place Q_n remain the same as that in the original DSPN model, while the firing rate of transition s_n is associated with the random switch $g_n(\mathbf{M})$ of the immediate transition d_n in the original DSPN. The resource sharing relationship of the original DSPN model is implicitly expressed in the marking-dependent firing rate of s_n as

$$\mu'_n = \mu_n g_n(\mathbf{M}). \tag{3.20}$$

By decomposition, the original multiuser system is represented by N subsystems, each of which consists of one SPN in Fig. 3.2a and one DSPN in Fig. 3.2b. Obviously, if each subsystem can be analyzed separately, the model decomposition can significantly reduce the size of the state space in the analysis and achieve better performance in computational complexity. Since each subsystem n in the decomposed DSPN model is almost the same as that defined for the single user system, a similar two-dimensional Markov model $(\mathcal{H}_{n,t}, \mathcal{Q}_{n,t})$ as described in Sect. 3.3.1 can be constructed for each subsystem. The only difference is that the firing rate of s_n becomes μ'_n instead of μ_n.

Unfortunately, the random switch $g_n(\mathbf{M})$ at the t-th time slot depends not only on its own marking, but also on the markings of all other subsystems. So such model decomposition is not "clean", i.e., there exist interactions among subsystems due to the marking-dependent firing rate μ'_n. The interactions belong to rate relation, which is a basic structure of near-independent model discussed in Sect. 2.2. Thus, in order to solve the n-th subsystem, the markings of all other subsystems have to be available at the same time. There are two difficulties arising from this situation: (1) the marking of a subsystem is equivalent to the states of a Markov chain, which is time-dependent and cannot be used/derived as the input/output parameters of any other subsystems; (2) the subsystems cannot be ordered, giving rise to cycles in the model solution process. Considering a two-user system as an example, each subsystem 1 or 2 needs the output of the other as input and neither can be first solved correctly. In order to solve these difficulties, two methods are introduced in the following two subsections. The first difficulty will be solved by using the steady-state probabilities of the markings instead of the instant markings as the subsystem input, and the second difficulty will be solved by fixed point iteration as discussed in Sect. 2.2. Simulation results shown in Sect. 3.4 indicate that such approximation is reasonable.

2) *Subsystem solution*:   Let $\boldsymbol{\pi}_n$ denote the steady-state probabilities of the embedded Markov chain $\{\mathcal{H}_{n,t}, \mathcal{Q}_{n,t}\}$ (i.e., the steady-state probabilities of the markings) for the n-th subsystem. Due to the interactions between the subsystems, the solution $\boldsymbol{\pi}_n$ for the n-th subsystem can be obtained only when the solutions of all the other subsystems $\{\boldsymbol{\pi}_i\}_{i=1,i\neq n}^{N}$ are known. Obviously, if the N subsystems are solved sequentially by index, the above requirement cannot be satisfied since only $\{\boldsymbol{\pi}_i\}_{i=1}^{n-1}$ are known when solving the n-th subsystem. In this subsection, we will derive the solution for the n-th subsystem under the assumption that $\{\boldsymbol{\pi}_i\}_{i=1,i\neq n}^{N}$ are given. We will leave the discussion of how to satisfy this assumption using the fixed point iteration method in the next subsection.

Given $\{\boldsymbol{\pi}_i\}_{i=1,i\neq n}^{N}$, we can approximate the random switch $g_n(\mathbf{M})$ with $\tilde{g}_n(\mathbf{M})$, which is a function of the steady-state probabilities of the markings. Notice that $g_n(\mathbf{M})$ indicates the probability that user n will be selected given a certain marking $\mathbf{M}$, while $\tilde{g}_n(\mathbf{M})$ indicates the long-run selection probability for user n. Based on $\tilde{g}_n(\mathbf{M})$, the firing rate of s_n can be determined as

$$\tilde{\mu}'_n = \begin{cases} \mu_n, & \text{with probability } \tilde{g}_n(\mathbf{M}), \\ 0, & \text{with probability } 1 - \tilde{g}_n(\mathbf{M}), \end{cases} \tag{3.21}$$

In the follows, the derivation of $\tilde{g}_n(\mathbf{M})$ for different scheduling algorithms is discussed.

- For the RR algorithm, the selection probability of user n can be expressed as

$$\tilde{g}_n(\mathbf{M}) = \sum_{U \in \mathcal{U}} \frac{1}{\|U\|} \prod_{i \in U, n \neq i} \mathbb{P}[M(Q_i) > 0] \prod_{j \in \overline{U}} \mathbb{P}[M(Q_j) = 0], \tag{3.22}$$

where $U = \{n, \dots\}$ is a set of users that *must* include user n and *may* include any other users, and $\mathcal{U}$ is the set that contains all possible sets of U.

- For the CA algorithm, the selection probability of user n depends on the server state. If the server is in the l-th state for user n, i.e. $M(H_{nl}) = 1$, we have $\mu_n = R_{n,l}$, and $\tilde{g}_n(\mathbf{M})$ can be expressed as

$$\tilde{g}_n(\mathbf{M}) = \sum_{U \in \mathcal{U}} \prod_{i \in U, n \neq i} \mathbb{P}[M(Q_i) > 0] \prod_{j \in \overline{U}} \mathbb{P}[M(Q_j) = 0] p_U(\mathbf{M}), \quad (3.23)$$

where

$$p_U(\mathbf{M}) = \sum_{V \in \mathcal{V}} \frac{1}{\|V\|} \prod_{i \in V, n \neq i} \mathbb{P}[M(H_{il}) = 1] \prod_{j \in U, j \in \overline{V}} \sum_{m=1}^{l-1} \mathbb{P}[M(H_{jm}) = 1],$$

$$(3.24)$$

$V \subseteq U$ and $\mathcal{V}$ is the set that contains all possible subsets of U.

- For the QA algorithm, the selection probability of user n depends on its queue length. If the queue length of user n is k, i.e. $M(Q_n) = k$, $\tilde{g}_n(\mathbf{M})$ can be expressed as

$$\tilde{g}_n(\mathbf{M}) = \sum_{U \in \mathcal{U}} \prod_{i \in U, n \neq i} \mathbb{P}[M(Q_i) > 0] \prod_{j \in \overline{U}} \mathbb{P}[M(Q_j) = 0] p_U(\mathbf{M}), \quad (3.25)$$

where

$$p_U(\mathbf{M}) = \sum_{V \in \mathcal{V}} \frac{1}{\|V\|} \prod_{i \in V, n \neq i} \mathbb{P}[M(Q_i) = k] \prod_{j \in U, j \in \overline{V}} \mathbb{P}[M(Q_j) < k].$$

$$(3.26)$$

- For the CQA algorithm, the selection probability of user n depends on both the server state and its queue length. If the server is in the l-th state and the queue length is k for user n, i.e. $M(H_{nl}) = 1$ and $M(Q_n) = k$, $\tilde{g}_n(\mathbf{M})$ can be expressed

$$\tilde{g}_n(\mathbf{M}) = \sum_{U \in \mathcal{U}} \prod_{i \in U, n \neq i} \mathbb{P}[M(Q_i) > 0] \prod_{j \in \overline{U}} \mathbb{P}[M(Q_j) = 0] p_U(\mathbf{M}), \quad (3.27)$$

where

$$p_U(\mathbf{M}) = \sum_{V \in \mathcal{V}} \frac{1}{\|V\|} \prod_{i \in V, n \neq i} \mathbb{P}[\mu_i M(Q_i) = \mu_{nl} k] \prod_{j \in U, j \in \overline{V}} \mathbb{P}[\mu_j M(Q_j) < \mu_{nl} k].$$

$$(3.28)$$

With $\tilde{g}_n(\mathbf{M})$, each subsystem n can be solved by following the similar way as in Sect. 3.3.1 with the following revisions for some system parameters.

- The computation of $p^n_{(l,k),(m,h)}$ should be changed to

$$p^n_{(l,k),(m,h)} = (v^{n,l}_{k,h}\tilde{g}_n(\mathbf{M}) + v^{n,0}_{k,h}(1 - \tilde{g}_n(\mathbf{M})))p^n_{l,m}, \qquad (3.29)$$

where $v^{n,l}_{k,h}$ and $v^{n,0}_{k,h}$ are determined according to (3.11) with $R_{n,0} = 0$.
- The mean throughput of user n becomes

$$\overline{T}_n = \sum_{l=1}^{L}\sum_{k=1}^{K} T^n_{l,k}\tilde{g}_n(\mathbf{M})\pi^n_{l,k}, \qquad (3.30)$$

where $T^n_{l,k}$ is derived according to (3.16). The total mean throughput of the multiuser system is

$$\overline{T} = \sum_{n=1}^{N}\overline{T}_n. \qquad (3.31)$$

- The dropping probability can be computed as (3.19). However, the value of $\mathbb{P}(B^n_{l,k} = b)$ equals

$$\mathbb{P}(B^n_{l,k} = b) = \mathbb{P}(A_{n,t} = K + b - \max[0, k - R_{n,l}\Delta T])\tilde{g}_n(M)$$
$$+\mathbb{P}(A_{n,t} = K + b - k)(1 - \tilde{g}_n(\mathbf{M})). \qquad (3.32)$$

3) *Fixed point iteration*: In the previous subsection, we assume that the solutions of all the other subsystems $\{\pi_i\}^N_{i=1,i\neq n}$ are already derived and can be used as input parameters when solving the n-th subsystem. However, this assumption is not true if the N subsystems are solved sequentially by index. In this subsection, fixed point iteration is used to deal with this problem.

Let $\{\pi_1, \ldots, \pi_N\}$ be the vector of *iteration variables* of the *fixed point equation*

$$\{\pi_1, \ldots, \pi_N\} = f(\{\pi_1, \ldots, \pi_N\}), \qquad (3.33)$$

where the function f is realized by solving the N subsystems successively with the subsystem solution method as described in the previous subsection. That is, the function f can be decomposed into N independent functions f_n, $n = 1, \ldots, N$, with f_n representing the solution of the n-th subsystem

$$\pi_n = f_n(\{\pi_1, \ldots, \pi_N\}).$$

Fig. 3.3 The z-th iteration
for the three-user scenario

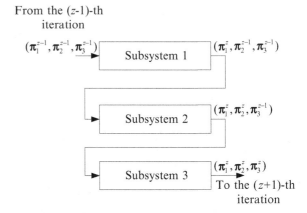

From the $(z-1)$-th
iteration

Obviously, the vector of steady-state distributions of the N subsystems $\{\pi_1, \ldots, \pi_N\}$ satisfies (3.33), which is referred to as the *fixed point* of this equation.

The fixed point can be derived by *successive substitution* in Sect. 2.2. Let the initial vector of iteration variables be $\{\pi_1^0, \ldots, \pi_N^0\}$. Each element of π_n^0 ($n = 1, \ldots, N$) can be set to an arbitrary value between 0 and 1. In the z-th iteration, we have

$$\{\pi_1^z, \ldots, \pi_N^z\} = f(\{\pi_1^{z-1}, \ldots, \pi_N^{z-1}\}), \tag{3.34}$$

where the iteration variables are determined by the function f based on the values of the last iteration, and the function f is realized by solving the N subsystems successively using the solution method as described above. Specifically, in solving the n-th subsystem in the z-th iteration, $\{\pi_1^z, \ldots, \pi_{n-1}^z, \pi_n^{z-1}, \ldots, \pi_N^{z-1}\}$ is used as the input to derive the value of π_n^z. After that, π_n^{z-1} is replaced by π_n^z as input in solving the rest of the subsystems (from the $(n+1)$-th to the N-th subsystems) during the z-th iteration. An example for the z-th iteration of a three-user system is illustrated in Fig. 3.3.

The iteration is terminated when the differences between the iteration variables of two successive iterations are less than a certain threshold value. The convergence of the fixed point iteration is proved in Appendix 2.

The computational complexity of the proposed analytical approach can be obtained as follows. Let Z represent the number of iterations for analysis to converge. Since in each iteration, the steady-state probabilities of the N subsystems have to be derived, the total time for the approach termination is $T = Z \times N \times T_{\text{sub}}$, where T_{sub} denotes the amount of time to solve each subsystem. T_{sub} depends on the state number of the embedded Markov chain for each subsystem, which equals $(K+1) \times L$. Note that the state number of the embedded Markov chain before decomposition equals $\big((K+1) \times L\big)^N$.

Therefore, the proposed analytical approach with decomposition and iteration techniques greatly reduces the computational complexity.

3.4 Numerical Results

In this section, both analytical and simulation results are presented to compare the performance of different scheduling algorithms. In the simulation, all users are assumed to have statistically identical wireless channels. The SNR thresholds and the corresponding transmission rates for each service process are defined in Table 3.1 [10]. If the instantaneous SNR is below -12.5 dB, the transmission rate is set to be 0. Accordingly, the FSMC model has 12 states in total. The carrier frequency f and the time slot duration ΔT are set to 1.9 GHz and 1.67 ms, respectively. The velocity v of the mobile users is assumed to be 3 km/h so that the Doppler frequency f_d^n becomes 5 Hz. We also let the mean SNR $\bar{\gamma}_n$ be 0 dB and the buffer size $K = 2{,}500$ bits. Notice that for other values of SNR, similar observations can be obtained.

In order to simplify the simulations by further reducing the state space of the Markov model, we redefine the data unit in the queueing system, or equivalently the token unit in the Petri net model, so that one data unit represents 50 bits. This approximation is acceptable since the amount of data that arrive to or depart from the system in one time slot usually includes a large number of bits. After redefinition, both the transmission rate $R_{n,l}$ and the buffer size K should be divided by 50, and the arrival rate and the performance metrics derived in the follows are in terms of data units. Note that the maximum queue length becomes 50 data units.

Figure 3.4 examines the accuracy of our analysis method described in Sect. 3.3. The stationary queue length distributions from both analytical and simulation results are compared, which are denoted by "Analysis" and "Simulation" in the figure,

Table 3.1 SNR threshold and rates

SNR threshold $\gamma_{n,l}$ (dB) $\geq$	Rates $R_{n,l}$ (Kbs)
-12.5	38.4
-9.5	76.8
-8.5	102.6
-6.5	153.6
-5.7	204.8
-4.0	307.2
-1.0	614.4
1.3	921.6
3.0	1228.8
7.2	1843.2
9.5	2457.6

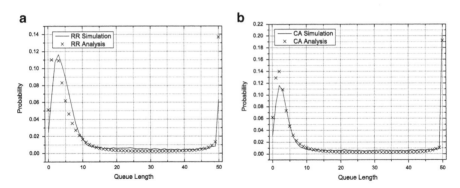

Fig. 3.4 Analytical and simulation results of queue length distribution for different schedulers. (**a**) Round robin algorithm, (**b**) channel-aware algorithm

respectively. In the simulation, the number of users is set to 2, and the mean data arrival rate per user is fixed at 1,500 data units per second. Each simulation runs for 20,000 time slots. Note that in the simulation, the analytical results converge after three iterations. In each iteration, the steady-state probabilities of the two embedded Markov chains for subsystems 1 and 2 are obtained, respectively. The state number of the Markov chain for each subsystem 1 or 2 equals 612, while the state number of the embedded Markov chain for directly modelling the whole system equals 374,544. This further demonstrates the computational complexity reduction by the proposed approach.

In Fig. 3.4, two subfigures (a) and (b) correspond to the RR and the CA algorithms, respectively. From the figures, it can be observed that, for both scheduling algorithms, the analytical results match well with those from the simulation, i.e., the proposed analytical method is accurate. The same observation is also true for the QA and the CQA algorithms, and the simulation and analytical results for these two algorithms are omitted. Since the buffer size K equals 50 data units and the arriving data is dropped if the buffer is full, the distribution of the queue length is truncated at 50. Since the queue length distribution at 50 represents the sum probability of the queue length equal to and larger than 50 if no buffer limitation exists, one peak at 50 is observed. In what follows, we will apply analytical results only to compare the performance of the four scheduling algorithms. The performance metrics include average queue length, mean throughput, average delay and dropping probability.

Figure 3.5 shows the performance comparison among the algorithms under different arrival rates (or traffic loads). The results are based on the fixed user number of 2 and the variant average data arrival rate per user from 1,000 to 40,000 data units per second. From the figure, it can be observed that (1) when the arrival rate is below a certain threshold (approximately 3,000–5,000 data units per second with respect to different performance metrics), the CA algorithm has the worst performance, while the QA algorithm achieves the best one; (2) when the arrival rate is higher than the threshold, the performance of the QA algorithm becomes the worst, while those of CA and CQA algorithms get very close and become the best;

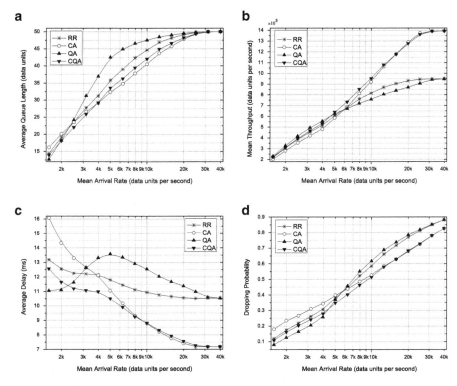

Fig. 3.5 Performance comparison under different arrival rates. (**a**) Average queue length (data units), (**b**) mean throughput (data units per second), (**c**) average delay (ms), (**d**) dropping probability

(3) the performance of the RR algorithm converges to that of the QA algorithm with the increase of the arrival rate. All these observations can be explained as follows. The CA algorithm tends to select a user with a better channel condition, while the QA algorithm favors a user with a larger queue length. When the traffic load is light, the instantaneous queue length is relatively short so that the actual transmission rate of a selected user n at any time slot t is mainly determined by its instantaneous queue length instead of its current channel condition. Therefore, in this case, the QA algorithm, which favors users with larger queue lengths, can maximize the transmission rate, while the CA algorithm performs worst due to the ignorance of queue length information. When the traffic load is heavy, on the other hand, the queue length is relatively large and the actual transmission rate of a selected user n at any time slot t is determined by its instantaneous channel condition. Therefore, the QA and the CA change their positions in system performance. Since the CQA algorithm considers both channel and queue states in user selection, it can balance the service rate and the queue length in both situations and thus achieve relatively good performance under all traffic conditions. When the traffic load is extremely

heavy, i.e., the queues of the users are saturated, the queue length information is not important any more in user selection, which results in the performance convergence of the CA/CQA and RR/QA algorithms. Also, note that in Fig. 3.5c, the average delay of QA algorithm is concave (increases before the arrival rate reaches 5,000 data units per second and decreases afterwards) while the average delay of others are monotonically decreasing with traffic load. This is because when the traffic load is light, the average queue length of the QA algorithm increases much faster than its mean throughput as shown in subfigures (a) and (b).

Figure 3.6 shows the performance of the scheduling algorithms under different numbers of users. The average data arrival rates per user are set at the low (1,500 data units per second) and high (5,000 data units per second) levels. The number of users varies from 2 to 10. In infinite backlog traffic model [11], it is well-known that the OS algorithms achieve larger throughput than the RR algorithm, or the scheduling gain, defined as the ratio between the throughput of an OS algorithm and the RR algorithm, is larger than 1. Furthermore, such scheduling gain increases with the number of users due to the improved multiuser diversity effect. However, Fig. 3.6a, b reveal that for dynamic data arrivals, the above observations only hold for high average arrival rate (5,000 data units per second). When the average arrival rate is low (1,500 data units per second), the scheduling gain of the CA algorithm over the RR algorithm becomes smaller than 1, and decreases with the increase of the number of users. A similar observation can be obtained for the average delay and the dropping probability, as well. When the average arrival rate is high, the performance of CA algorithm deteriorates more slowly in terms of the average delay and the dropping probability than the RR algorithm as the number of users increases. But, when the average arrival rate is low, an opposite results can be observed. The analytical results for these two performance metrics are omitted here due to space limitation. The faster performance degradation of the CA algorithm under light traffic load results from the facts that (1) the multiuser diversity gain has little effects on the transmission rate, which is dominated by the queue length; and (2) such negative effects are enlarged in terms of delay and dropping probability when more users are waiting for transmission.

3.5 Summary

In this chapter, a general framework for analyzing performance of the wireless schedulers in multiuser systems has been discussed. The system behavior is formulated by an M/MMDP/1/K queueing model and the approximation for multiple performance metrics is derived with low computational complexity by applying the decomposition and the iteration techniques from SPNs. Based on this framework, four classes of typical wireless schedulers, which are referred to as round robin, channel-aware, queue-aware and channel/queue-aware, are analyzed and compared in terms of average queue length, mean throughput, average delay and dropping probability. The analysis shows that

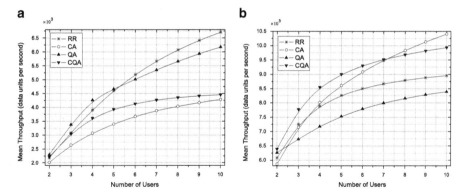

Fig. 3.6 Performance comparison under different number of users. (**a**) Mean throughput—arrival rate 1.5 Kps, (**b**) mean throughput—arrival rate 5 Kps

- while in heavy traffic regime the channel-aware scheduler indeed outperforms the round robin scheduler, this is not true when the traffic load is light;
- the performance of the channel/queue-aware scheduler is better than that of the channel-aware scheduler in the light traffic regime, and converges to the latter with the increase of the traffic load; and
- the ratio of the throughput between the channel-aware scheduler and the round robin scheduler, which is commonly referred to as the scheduling gain, decreases with the increase of the number of users when the traffic load per user is light.

Appendix 1: Determination of $p_{l,m}^n$ in Rayleigh Fading Channel

For Rayleigh fading channel, $p_{l,m}^n$ can be determined as follows [12]. Assume the state transitions of the FSMC happen only between adjacent states, i.e.

$$p_{l,m}^n = 0, \quad |l - m| \geq 2. \tag{3.35}$$

Let $\gamma_{n,l}$, $(l = 1, \ldots, L - 1)$, denotes the SNR threshold value between the l-th and $(l + 1)$-th states of the FSMC model for user n. The adjacent-state transition probability can be calculated as

$$p_{l,l+1}^n = \frac{\chi(\gamma_{n,l+1})\Delta T}{\pi_{n,l}}, \quad l = 1, \ldots, L - 1, \tag{3.36}$$

$$p_{l,l-1}^n = \frac{\chi(\gamma_{n,l})\Delta T}{\pi_{n,l}}, \quad l = 2, \ldots, L. \tag{3.37}$$

Here, $\chi(\gamma_n)$ denotes the level cross rate at an instantaneous SNR γ_n and is given by

$$\chi(\gamma_n) = \sqrt{\frac{2\pi\gamma_n}{\overline{\gamma}}}\, f_d^n \exp(-\frac{\gamma_n}{\overline{\gamma}_n}), \tag{3.38}$$

where f_d^n denotes the mobility-induced Doppler spread, $\overline{\gamma}_n = \mathbb{E}[\gamma_n]$ is the average received SNR, and $\pi_{n,l}(l \in \mathcal{L})$ denotes the stationary probability that the FSMC is in state l given by

$$\pi_{n,l} = \exp(\gamma_{n,l}/\overline{\gamma}_n) - \exp(\gamma_{n,l+1}/\overline{\gamma}_n). \tag{3.39}$$

Finally, $p_{l,l}^n$ can be derived from the normalizing condition $\sum_{m=1}^{L} p_{l,m}^n = 1$ as

$$p_{l,l}^n = \begin{cases} 1 - p_{l,l+1}^n - p_{l,l-1}^n, & (l = 2, \ldots, L-1), \\ 1 - p_{l,l+1}^n, & (l = 1), \\ 1 - p_{l,l-1}^n, & (l = L). \end{cases} \tag{3.40}$$

Appendix 2: Convergence of the Fixed Point Iteration

According to Sect. 2.2, in order to prove the convergence of the fixed point iteration for the decomposed DSPN model as described in (3.34), it is sufficient to show that the following lemma is true.

Lemma 3.1. *The embedded Markov chain $(\mathcal{H}_{n,t}, \mathcal{Q}_{n,t})$ for the n-th subsystem is irreducible, if $K \leq R_{n,L}\Delta T$.*

Proof. It has been proved in [7] that the Markov chain for the single user system is irreducible. Similarly, we prove Lemma 3.1 by showing that for each transition from state (l, k) to (m, h), there exists a multi-transition path $(l, k) \rightarrow (l^*, k) \rightarrow (l^*, h) \rightarrow (m, h)$ with non-zero probability, where $R_{n,l^*}\Delta T \geq k$. Since $K \leq R_{n,L}\Delta T$, there always exists such l^* that satisfies this condition.

Since the FSMC model is irreducible, we have that p_{l,l^*}^n, p_{l^*,l^*}^n and $p_{l^*,m}^n$ are all positive. Now we shall verify the following inequalities:

(a) $v_{k,k}^{n,l} \tilde{g}_n(\mathbf{M}) + v_{k,k}^{n,0}(1 - \tilde{g}_n(\mathbf{M})) > 0$;

(b) $v_{k,h}^{n,l^*} \tilde{g}_n(\mathbf{M}) + v_{k,h}^{n,0}(1 - \tilde{g}_n(\mathbf{M})) > 0$;

(c) $v_{h,h}^{n,l^*} \tilde{g}_n(\mathbf{M}) + v_{h,h}^{n,0}(1 - \tilde{g}_n(\mathbf{M})) > 0$.

For inequality 1), since $A_{n,t} = k - \max[0, k - R_{n,l}\Delta T] \geq 0$ ($R_{n,0} = 0$ included), we have $v_{k,k}^{n,l} > 0$ and $v_{k,k}^{n,0} > 0$. Therefore, inequality 1) is true with $\tilde{g}_n(\mathbf{M}) \in [0, 1]$. The proof of inequality 3) is similar.

For inequality 2), since $A_{n,t} = h - \max[0, k - R_{n,l*} \Delta T] \geq 0$, we have $v_{k,h}^{n,l*} > 0$, where $R_{n,l*} \Delta T \geq k$. Now consider both the cases when the value of k is zero or not. If $k = 0$, we have $\tilde{g}_n(\mathbf{M}) = 0$ and $v_{k,h}^{n,0} > 0$; otherwise, if $k > 0$, we have $\tilde{g}_n(\mathbf{M}) > 0$ and $v_{k,h}^{n,0} \geq 0$. Thus, the inequality 2) is true under both cases.

According to (3.29), we have $p_{(l,k),(l*,k)}^n > 0$, $p_{(l*,k),(l*,h)}^n > 0$ and $p_{(l*,h),(m,h)}^n > 0$, where $R_{n,l*} \Delta T \geq k$, which prove the existence of the multi-transition path with non-zero probability for each transition from state (l, k) to (m, h).

References

1. 3GPP2 C.S0024 Version 4.0 (2002) CDMA 2000 High Rate Packet Data Air Interface Specification
2. M. Andrews (2004) Instability of the proportional fair scheduling algorithm for HDR. IEEE Trans. Wireless Commun 3(5):1422–1426
3. M. Andrews et al (2004) Scheduling in a queueing system with asynchronously varying service rate. Probability in the Engineering and Informational Sciences 18:191–217
4. D. Wu, R. Negi (2005) Utilizing multiuser diversity for efficient support of quality of service over a fading channel. IEEE Trans. Veh. Technol 54(3):1198–1206
5. F. Ishizaki, G.U. Hwang (2007) Queueing delay analysis for packet schedulers with/without multiuser diversity over a fading channel. IEEE Trans. Veh. Technol 56(5):3220–3227
6. S. Shakkottai (2008) Effective capacity and QoS for wireless scheduling. IEEE Trans. Automatic Control 53(3):749–761
7. Q. Liu, S. Zhou, G. B. Giannakis (2005) Queueing with adaptive modulation and coding over wireless links: cross-layer analysis and design. IEEE Trans. Wireless Commun 50(3):1142–1153
8. G. Bolch and C. Bruzsa (1995) Modeling and simulation of Markov modulated multiprocessor systems with Petri nets. Paper presented at the 7th European Simualtion Symprosium, University of Erlangen(Germany), 1995
9. T. Yang, D. H. K. Tsang (1995) A novel approach to estimating the cell loss probability in an ATM multiplexer loaded with homogeneous on-off sources. IEEE Trans. Commun 43(1):117–126
10. P. Bender et al (2000) CDMA/HDR: a bandwidth-efficient high-speed wireless data service for nomadic users. IEEE Commun. Mag 38(7):70–77
11. S. C. Borst (2005) User-level performance of channel-aware scheduling algorithms in wireless data networks. IEEE/ACM Trans. Networking 13(3):636–647
12. Q. Zhang, S. A. Kassam (1999) Finite-state Markov model for Rayleigh fading channels. IEEE Trans. Commun 47(11):1688–1692

Chapter 4
Performance Analysis of Device-to-Device Communications with Dynamic Interference Using SPNs

In Chap. 3, we have adopted the model decomposition and iteration techniques in the SPNs to analyze the performance of opportunistic schedulers. In this chapter, we use this approximation technique to study the performance of D2D communications with dynamic interference [1]. In specific, we analyze the performance of frequency reuse among D2D links with dynamic data arrival setting. We first consider the arrival and departure processes of packets in a non-saturated buffer, which result in varying interference on a link based on the change of its backlogged state. The packet-level system behavior is then represented by a coupled processor queuing model, where the service rate varies with time due to both the fast fading and the dynamic interference effects. In order to analyze the queuing model, we formulate it as a DTMC and compute its steady-state distribution. Since the state space of the DTMC grows exponentially with the number of D2D links, we use the model decomposition and some iteration techniques in SPNs to derive its approximate steady state solution, which is used to obtain the approximate performance metrics of the D2D communications in terms of average queue length, mean throughput, average packet delay and packet dropping probability of each link. Simulations are performed to verify the analytical results under different traffic loads and interference conditions. Note that Chaps. 3 and 4 show two typical resource sharing paradigms in wireless networks—orthogonal sharing by user scheduling and non-orthogonal sharing by frequency reuse, which induce different interactions among the decomposed submodels.

4.1 Packet Level Performance Analysis of D2D Communications

D2D communications are commonly referred to as the type of the technologies that enable devices to communicate directly with each other without the infrastructure, e.g, access points or base stations. Bluetooth and WiFi-Direct are two most popular

© The Author(s) 2015
L. Lei et al., *Stochastic Petri Nets for Wireless Networks*, SpringerBriefs
in Electrical and Computer Engineering, DOI 10.1007/978-3-319-16883-8_4

D2D techniques in the market, bothing working in the unlicensed 2.4 GHz ISM bands. Cellular networks, on the other hand, do not support direct over-the-air communications between user devices. However, with the emergence of context-aware applications and the accelerating growth of Machine-to-Machine (M2M) applications, D2D function plays a more and more important role since it facilitates the discovery of geographically close devices and reduce the communication cost between these devices. To seize the emerging market that requires D2D function, the mobile operators and vendors are exploring the possibilities of introducing D2D function in the cellular networks to develop the network assisted D2D communications technologies [2–5]. The most significant difference between the cellular network assisted D2D communications and the traditional D2D technologies such as WiFi-direct is that the former works in the licensed band of cellular networks with more controllable interference and the base station or the network can assist the D2D user equipments (UEs) in various functions, such as new peer discovery methods, physical layer procedures, and radio resource management algorithms.

In the emerging new cellular networks such as 3GPP LTE, orthogonal time-frequency resources are allocated to different users within a cell to eliminate intracell interference. The introduction of D2D function may bring two categories of potential intracell interference into cellular networks: interference between different D2D users and interference between a cellular user and one or multiple D2D users. The former category of interference arises when the radio resources are reused by multiple D2D users, while the latter arises when the radio resources allocated to a cellular user are reused by one or more D2D users. The latter category of interference can be avoided by statically or semi-statically allocating a group of dedicated resources to all the D2D users at the cost of reduced spectrum efficiency. Most of the existing work on D2D communications focus on the design of optimized resource management algorithms using a static interference model, where each D2D link or cellular link is assumed to be saturated with infinite backlogs and constantly cause interference to the other D2D links or cellular links [6–12]. The base station can either centrally determine the radio resource allocation of D2D connections along with the cellular connections [5, 6] or let the users perform distributed resource allocation by themselves [4, 7]. In this chapter, we focus on the first category of interference (i.e., interference between D2D links) with centralized resource allocation and study the performance of D2D communications using a dynamic interference model, where the finite amount of data arrives to the links at each time slot. Thus, the links do not always have data to transmit and cause interference to the other links. In order to focus on the dynamic interference, we consider the Full Reuse (FR) resource sharing strategy in this chapter, where all the available resources are reused by all the D2D links with non-empty queues in each time slot.

In the queuing model for such system, the service rate at each queue varies with time due to two factors: the fast fading effect of the wireless channel and the dynamic arrival and departure of packets which results in the dynamic variation of interference from a link when its status changes from busy to idle or vice versa. When only the second factor is considered, the system can be modeled as a coupled-processor server, where the service rates at each queue vary over time as governed by the backlogged state of the other queues. The complex interaction between the various queues renders an exact analysis intractable in general and steady-state queue length distributions are known only for exponentially distributed service in two-queue systems [13–15]. In [16, 17], the flow-level performance in wireless networks with multiple base stations is investigated, which can be formulated as a coupled-processor model in the single user class scenario. Due to the complex nature of the coupled-processor model, Bonald et al. [16] derives bounds and approximations for the key performance metrics by assuming maximum and minimum interference in neighbours of the reference cell and [17] studies the performance gains of intercell scheduling in a two-cell network and a simplified symmetric network, respectively. In [18], the upper and lower bounds on the moments of the queue length of the coupled-processor model are obtained by formulating a moments problem and solving a semidefinite relaxation of the original problem. This analysis method is applied in [19] to study the impact of user association policies on flow-level performance in interference-limited wireless networks. Although the semidefinite relaxation method can be applied to study the coupled-processor model with more than two queues, the size of the formulated optimization problem scales exponentially as the number of queues increases.

In this chapter, we consider both the fast fading and dynamic interference effects to better depict the variations of the service rate in the practical D2D system. The FSMC model with respect to the SINR is constructed for each link, which not only captures the fast fading effect of the wireless channel as in the traditional FSMC model based on SNR partition [20, 21], but also considers the dynamic variation of interference due to the changing backlogged states of the other active links. Based on the above FSMC model, we formulate a coupled processor queuing model for such a system with time-varying service rates and propose an analytical method to derive the state transition probability matrix and steady-state distribution of the underlying Markov process. However, the scalability of proposed analytical method with large number of links is limited by the exponentially increasing state space of the Markov process. Therefore, we use the model decomposition and iteration approach in SPN which has been discussed in Sect. 2.2 to deal with the coupling between the service rates of the different links. Specifically, we formulate the SPN model for the above queuing system with multiple D2D links and decompose it into multiple "near-independent" subnets, where each subnet corresponds to a queuing system with a single link and is solved separately. Because the subnets are correlated, after solving the problem in each subnet to get its steady-state statistics, the distributions are exported to other subnets to derive their approximate state transition probability matrices and steady-state distributions, and this is conducted iteratively. Finally, performance metrics such as throughput and packet dropping probability can be obtained from the steady-state distribution of the Markov process. Note that although both Chaps. 3 and 4 use the model decomposition and iteration approach to analyze the performance of wireless networks, the interaction between different subsystems is due to scheduling and interference, respectively.

4.2   The Coupled Processor Queueing System

Consider a cellular wireless network with D2D communication capability. Figure 4.1a illustrates the case where two D2D links share radio resources with each other. A D2D link consists of a source D2D User Equipment (UE) transmitting to its destination D2D UE. Potential interfering link exists for D2D link 1 (resp. D2D link 2) from the transmitter of source D2D UE2 (resp. source D2D UE1) to the receiver of destination D2D UE1 (resp. destination D2D UE2). Since there are two categories of links, i.e., D2D links and potential interfering links, all links mentioned are referred to the D2D links by default in the rest of the chapter. We consider that each source D2D UE maintains a queue with finite capacity to buffer the dynamically arriving data. An interfering link is 'potential' since it only exists when the queue of the source D2D UE associated with its transmitter is non-empty. The transmission rate of each link is determined by its own SINR, which varies with time due to the fast fading effects of its own wireless channel, and also the fast fading effects and changing on-off status of its potential interfering link. Moreover, the transmission rates of different links vary asynchronously over time. Therefore, Fig. 4.1a can be modeled as a queuing system as illustrated in Fig. 4.1b, where there are two queues and each one is served by a private single-server, whose service rate is determined by the transmission rate of the corresponding link.

Let $\mathcal{D} = 1, \ldots, D$ denote the set of non-overlapping links. Each link maintains a queue at the source D2D UE, and each queue has a finite capacity of $K < \infty$ packets, where packets are assumed to be of the same size B bits. For each queue i, packets arrives according to Poisson distribution with average rate λ_i packets/s. The transmission in the time is slot-by-slot based and each slot has an equal length ΔT. In each time slot, the resource can be allocated to one or more links, depending on the resource sharing and scheduling strategies. Although there are various resource

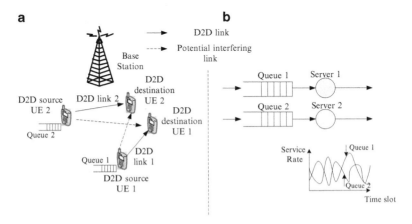

Fig. 4.1 Cellular wireless networks with D2D communications capability (**a**) resource sharing between two D2D links; (**b**) queuing model of (**a**)

sharing strategies between the links, there are two extreme cases which incur the maximum and minimum interference, respectively:

- FR: The links reuse all the available resources, causing interference to each other. However, the links get the largest amount of resources to use.
- OS: The links use orthogonal resources with each other where no interference exists. However, each link gets the least amount of resources to use.

We focus on the performance of FR strategy with dynamic interference. This resource sharing strategy is not a practical one, since it may cause excessive amount of interference between D2D links. However, it is an extreme case which incurs the maximum interference but achieves the best frequency reuse. On the other hand, the queuing performance of the other extreme case, i.e., the OS strategy, can be analyzed using the method in our earlier work [22]. Since the practical resource allocation strategies try to achieve the best tradeoff between frequency reuse and interference control, the performance of the two extreme cases provide lower bounds for the performance of practical resource allocation strategies. Moreover, the packet-level performance evaluation method of other practical resource allocation methods can be developed based on those of FR and OS strategies.

Assume that the instantaneous channel gains of the D transmitters Tx_i on the source D2D UEs of links $i \in \mathcal{D}$ with the D receivers Rx_j on the destination D2D UEs of links $j \in \mathcal{D}$ remain constant within a time slot, the value of which at time slot t can be represented by a D-by-D channel gain matrix $\mathbf{G}_t$, where item $G_{ij,t}$ denotes the channel gain between the transmitter Tx_i of link i and the receiver Rx_j of link j. The channel gain matrices $\mathbf{G}_t$ and $\mathbf{G}_{t'}$ in different time slots $t \neq t'$ can be different due to the fast fading effects of the wireless channel. Let $\mathcal{I}_i := \{I_{ji}\}_{j \in \mathcal{D} \setminus \{i\}}$ denote the set of potential interfering links of link i, where I_{ji} is the potential interfering link from the transmitter of link $j \in \mathcal{D} \setminus \{i\}$ to the receiver of link i. Note that the channel gain of link i is $G_{ii,t}$, while the channel gains of set of potential interfering links $\mathcal{I}_i$ are $\{G_{ji,t}\}_{j \in \mathcal{D} \setminus \{i\}}$. Let $\mathbf{P} = \{P_i\}_{i=1,\dots,D}$ be the D-by-1 power matrix that determines the transmission power of every link i.

We consider that a link does not always have data to transmit, and its transmitter first examines whether the queue is empty or not at the beginning of every time slot t. Only when the queue is non-empty shall it move the packets out of the queue for transmission and thus cause interference to the other links. We consider the transmission capability of link i during time slot t as $\lfloor \frac{r_{i,t} \Delta T}{B} \rfloor$, where $r_{i,t}$ is its instantaneous data rate in terms of bits/s and $\lfloor \cdot \rfloor$ is the integer no bigger than $\cdot$. Here, we assume that if a packet could not be transmitted completely due to time-slot expiration at the end of time-slot t, which can be foreseen at the beginning of this time slot, the entire packet is not transmitted in this time slot but at the next time-slot $(t + 1)$. Although a packet can be truncated for transmission in a realistic scenario, for example, by the RLC protocol in the LTE systems, assuming the realistic scenario will make the queue length a real number instead of an integer, which will result in the infinite state space of the Markov chain. Therefore, we make this approximation in our analysis, while keeping the approximation under control by adjust the packet size B. This approximation has also been used in existing work in literature[21, 23], where the service rate is given in terms of packets/s instead of bits/s. If the number of packets in the queue of link i at the beginning of time slot t is less than its transmission capability during time slot t, padding bits shall be transmitted along with the data according to the LTE standard. Arriving packets are placed in the queue throughout the time slot t and can only be transmitted during the next time slot $t + 1$. If the queue length reaches the buffer capacity K, the subsequent arriving packets will be dropped. Let $\vec{Q}_t = \{Q_{i,t}\}_{i \in \mathcal{D}}$ denote the queue length of every link i in terms of packets at the beginning of time slot t, and $A_{i,t}$ denote the number of packets arrived to link i during the time slot t, which is a Poisson distributed stationary process with mean $\lambda_i \Delta T$. According to the above assumption, the queuing process evolves following

$$Q_{i,t+1} = \min[K, \max[0, Q_{i,t} - \lfloor \frac{r_{i,t} \Delta T}{B} \rfloor] + A_{i,t}]. \tag{4.1}$$

From the above discussion, the SINR of every link i depends on the subset of other links in the system with non-zero queue length at the beginning of time slot t, which is denoted as $\mathcal{J}_{i,t}$. Let $\vec{\Theta}_t = \{\Theta_{i,t}\}_{i \in \mathcal{D}}$, where $\Theta_{i,t} = \mathbf{1}(Q_{i,t} > 0)$ denotes the queue status (empty or not) of the link i at the beginning of time slot t. Note that $\vec{\Theta}_t$ can take 2^D possible values $\theta_v, v = 1 \dots, 2^D$. Therefore, we have $\mathcal{J}_{i,t} := \{j \in \mathcal{D}\setminus\{i\} : \Theta_{j,t} = 1\}$. The SINR value for each link i at time slot t is given by the following formula:

$$SINR_{i,t} = \frac{P_i G_{ii,t}}{N_i + \sum_{j \in \mathcal{D}\setminus\{i\}} P_j G_{ji,t} \Theta_{j,t}} = \frac{\gamma_{ii,t}}{1 + \sum_{j \in \mathcal{D}\setminus\{i\}} \gamma_{ji,t} \Theta_{j,t}}, \tag{4.2}$$

where $\gamma_{ii,t} := \frac{P_i G_{ii,t}}{N_i}$ is the SNR value of link i, and N_i is the noise power on link i. Similarly, $\gamma_{ji,t} := \frac{P_j G_{ji,t}}{N_i}$, $j \in \mathcal{D}\setminus\{i\}$ can be referred to as the 'virtual SNR' value of the link I_{ji}, where it is "virtual" since I_{ji} is an interference link instead of a link and it is in fact the 'interference to noise ratio' considering the physical meaning.

The corresponding instantaneous data rate $r_{i,t}$ is a function of $SINR_{i,t}$. In this chapter, we assume that AMC is used, where the SINR values are divided into L non-overlapping consecutive regions and if the SINR value $SINR_{i,t}$ of link i falls within the l-th region $[\chi_{l-1}, \chi_l)$, the corresponding data rate $r_{i,t}$ of link i is a fixed value R_l, i.e., $r_{i,t} = R_l$, if $SINR_{i,t} \in [\chi_{l-1}, \chi_l)$. Table 4.1 gives the AMC scheme in 3GPP LTE systems where $L = 16$ [24]. As an example, $l = 2$ if $SINR_{i,t} \in [-4.46\,\text{dB}, -3.75\,\text{dB})$, and $R_2 = 213.3$ Kbs. Since the AMC function can select the appropriate coding and modulation schemes according to the instantaneous SINR of the wireless channel guaranteeing that the packet error rate is above an acceptable value, we do not consider the transmission errors. Although it is an interesting and challenging research problem to study the transmission errors due to factors such as imperfect Channel State Information (CSI), it is outside the scope of this chapter.

Table 4.1 SINR threshold and rates

Channel state index l	SINR threshold $\chi_{(l-1)}$ (dB) $\geq$	Rates R_l (Kbs)
2	−4.46	213.3
3	−3.75	328.2
4	−2.55	527.8
5	−1.15	842.2
6	1.75	1227.8
7	3.65	1646.1
8	5.2	2067.2
9	6.1	2679.7
10	7.55	3368.8
11	10.85	3822.7
12	11.55	4651.2
13	12.75	5463.2
14	14.55	6332.8
15	18.15	7161.3
16	19.25	7776.6

In order to achieve AMC, we assume that the BS has knowledge of the channel gain matrix $\mathbf{G}_t$ and the queue status $\vec{\Theta}_t$ of all the D2D links at each time slot t, so that it can determine the modulation and coding schemes for each D2D link with non-empty queues and inform the source and destination D2D UEs with downlink control signaling. Since network assisted D2D communications have not been standardized in 3GPP, there is currently no specific signaling protocols for resource allocations of D2D connections. In [2], we have proposed a candidate signaling procedure.

Since the SINR value $SINR_{i,t}$ of any link $i \in \mathcal{D}$ can be derived according to (4.2), the wireless channel for each link i can be modeled as a FSMC with total L states, where $H_{i,t}$ represents the channel state of the FSMC model of link i at time slot t. Each state of FSMC corresponds to one non-overlapping consecutive SINR region and a fixed transmission rate determined by the AMC algorithm. From (4.2), it can be seen that the SINR value $SINR_{i,t}$ and thus the channel state $H_{i,t}$ of link i depends on the SNR value $\gamma_{ii,t}$ of link i and the 'virtual SNR' values $\gamma_{ji,t}$ of its interfering links I_{ji}, $j \in \mathcal{D}\backslash\{i\}$, and also the queue status $\Theta_{j,t}$ of the links $j \in \mathcal{D}\backslash\{i\}$. For any link $i \in \mathcal{D}$, since both the (virtual) SNR values $\gamma_{ji,t}$, $j \in \mathcal{D}$ and the queue status $\Theta_{j,t}$, $j \in \mathcal{D}\backslash\{i\}$ remain constant within a time slot t, the SINR value $SINR_{i,t}$ also remains constant within a time slot.

There has been a lot of research on the finite state Markov modeling of wireless fading channels, where interference is not considered. Compared with these work, our FSMC has two additional complicating factors: (1) the fading of the potential interfering links of link i; (2) the variation of the set of interfering links of link i due to changing backlog status of the other links. For factor (1), the channel gain of an interfering link in time slot t can be considered as only dependent on its channel gain in time slot $t - 1$ due to the time correlation as assumed in the existing first-order FSMC models. For factor (2), the queue length of a link at time slot t only depends on its queue length at time slot $t - 1$ and the channel state $H_{i,t-1}$ at time slot $t - 1$, as will be discussed later and calculated in (4.4). Therefore, the channel state $H_{i,t}$ of link i at time slot t only depends on the channel gains of link i and its potential interfering links at time slot $t - 1$ and the queue length of the other links at time slot $t - 1$, which obeys the Markovian property.

The above D2D communications system can be formulated by a coupled processor queuing model as follows. The defined queuing system consists of a finite number, D, of queues indexed by $i = 1, 2, \ldots, D$, each of which has a server i corresponding to a D2D link i. For any i, there is a Poisson distributed packet arrival process with mean $\lambda_i \Delta T$ fixed length packets of B bits, and a finite and discrete-time Markov chain $H_{i,t}$ with total L states representing the evolution of the channel states of D2D link i. Associated with the l-th ($l \in \{1, \ldots, L\}$) state of the FSMC model of any link i is a fixed service rate R_l bits/s of server i, which is a non-negative integer and the same for all the links. Since the wireless channels vary with time asynchronously for different links, the transitions of the channel states are link dependent and the channel states $H_{i,t}$ and $H_{j,t}$ of any two different links $i \neq j$ at time slot t are not necessarily the same. If at time slot t the queue i is non-empty with $H_{i,t}$ in the l-th state, the queue i is served at a deterministic rate R_l, i.e., the queue is served according to an L-state MMDP. For any link $i \in \mathcal{D}$, since $SINR_{i,t}$ depends on $\{\Theta_{j,t}\}_{j \in \mathcal{D} \setminus \{i\}}$ and thus $\{Q_{j,t}\}_{j \in \mathcal{D} \setminus \{i\}}$, the state of $H_{i,t}$ depends on the set of queues in the system with non-zero queue length, which corresponds to a coupled processor server.

4.3 Steady-State Solution of the Queueing System

Let $\vec{H}_t := \{H_{i,t}\}_{i \in \mathcal{D}}$ denote the channel states for every link at time slot t. Let $\vec{Q}_t$ as defined in Sect. 4.2 represent the queue states for every link at time slot t. The $(2 \times D)$-dimensional DTMC $\{(\vec{H}_t, \vec{Q}_t), t = 0, 1, \ldots\}$ can be used to represent the system behavior of the above queuing system. The state number of the DTMC is $((K + 1) \times L)^D$, which grows exponentially with the increasing number D of D2D links. In this section, we focus on the derivation method of the exact transition probabilities and steady-state solution of the DTMC and leave the state space explosion problem to the next section, where the model decomposition and iteration method in SPN is used to decompose the DTMC with $((K + 1) \times L)^D$ number of states to D DTMCs each with $(K + 1) \times L$ number of states. Based on the exact method discussed in this section, the approximate transition probabilities and steady-state solution of the decomposed DTMCs can be derived.

Fig. 4.2 The computation of $p_{(\vec{l},\vec{k})}^{(\vec{n},\vec{h})}$, which mainly consists of two components. The first component $p_{(\vec{l},\vec{k})}^{\vec{h}}$ represents the transition probability of the queue state, and the second component $p_{(\vec{l},\vec{k},\vec{h})}^{\vec{n}}$ represents the transition probability of the server state

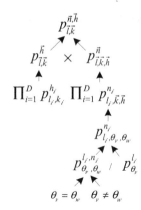

Let $p_{(\vec{l},\vec{k})}^{(\vec{n},\vec{h})}$ be the transition probability from state $(\vec{l},\vec{k})$ to state $(\vec{n},\vec{h})$ of the Markov chain, where $\vec{l} := \{l_i\}_{i\in\mathcal{D}}$, $\vec{k} := \{k_i\}_{i\in\mathcal{D}}$, $\vec{n} := \{n_i\}_{i\in\mathcal{D}}$ and $\vec{h} := \{h_i\}_{i\in\mathcal{D}}$. Note that $l_i, n_i \in \{1,\ldots,L\}$, and $k_i, h_i \in \{0,\ldots,K\}$. We can first decompose $p_{(\vec{l},\vec{k})}^{(\vec{n},\vec{h})}$ into two components as

$$p_{(\vec{l},\vec{k})}^{(\vec{n},\vec{h})} = \text{Pr.}\{\vec{Q}_{t+1} = \vec{h}|\vec{H}_t = \vec{l}, \vec{Q}_t = \vec{k}\} \times \text{Pr.}\{\vec{H}_{t+1} = \vec{n}|\vec{H}_t = \vec{l},$$

$$\vec{Q}_t = \vec{k}, \vec{Q}_{t+1} = \vec{h}\} = p_{(\vec{l},\vec{k})}^{\vec{h}} p_{(\vec{l},\vec{k},\vec{h})}^{\vec{n}}, \tag{4.3}$$

where the first component $p_{(\vec{l},\vec{k})}^{\vec{h}}$ is the transition probability of the queue state from $\vec{Q}_t = \vec{k}$ to $\vec{Q}_{t+1} = \vec{h}$, given the channel state $\vec{H}_t = \vec{l}$; and the second component $p_{(\vec{l},\vec{k},\vec{h})}^{\vec{n}}$ is the transition probability of the channel state from $\vec{H}_t = \vec{l}$ to $\vec{H}_{t+1} = \vec{n}$, given the queue states $\vec{Q}_t = \vec{k}$ and $\vec{Q}_{t+1} = \vec{h}$.

In the rest of Sect. 4.3, we will first discuss the computation method of $p_{(\vec{l},\vec{k})}^{\vec{h}}$ and $p_{(\vec{l},\vec{k},\vec{h})}^{\vec{n}}$, respectively, to get the state transition probability $p_{(\vec{l},\vec{k})}^{(\vec{n},\vec{h})}$ of the Markov chain of $\{(\vec{H}_t, \vec{Q}_t), t = 0, 1, \ldots\}$ according to (4.3). The framework of computing $p_{(\vec{l},\vec{k})}^{(\vec{n},\vec{h})}$ in the following part is illustrated in Fig. 4.2. Then, we will derive the steady-state distribution of the Markov chain from its state transition probability matrix and prove that the steady-state distribution exists and is unique under certain conditions in Theorem 4.3.

4.3.1 Transition Probability of the Queue State

According to (4.1) and given $r_{i,t} = R_{l_i}$, we have

$$p_{l_i,k_i}^{h_i} = \text{Pr.}\{Q_{i,t+1} = h_i | H_{i,t} = l_i, Q_{i,t} = k_i\} \tag{4.4}$$

$$= \begin{cases} \text{Pr.}(A_{i,t} = h_i - k_i + \eta_i) & \text{if } k_i > \eta_i, h_i \neq K, \\ \text{Pr.}(A_{i,t} = h_i) & \text{if } k_i \leq \eta_i, h_i \neq K, \\ \text{Pr.}(A_{i,t} \geq K - k_i + \eta_i) & \text{if } k_i > \eta_i, h_i = K, \\ \text{Pr.}(A_{i,t} \geq K) & \text{if } k_i \leq \eta_i, h_i = K, \end{cases}$$

where $\eta_i = \lfloor \frac{R_{l_i} \Delta T}{B} \rfloor$, and $\text{Pr.}(A_{i,t} = a) = \frac{(\lambda_i \Delta T)^a}{a!} e^{-\lambda_i \Delta T}$ due to Poisson assumptions.

Since the transition probability $p_{l_i,k_i}^{h_i}$ of each link i depends only on its own server and queue states, we have

$$p_{(\vec{l},\vec{k})}^{\vec{h}} = \prod_{i=1}^{D} p_{l_i,k_i}^{h_i}. \tag{4.5}$$

4.3.2 Transition Probability of the Server State

According to (4.2), the value of $SINR_{i,t}$ is determined by the (virtual) SNR vector $\vec{\gamma}_{i,t} := \{\gamma_{ji,t}\}_{j \in \mathcal{D}}$ and the queue status vector $\{\Theta_{j,t}\}_{j \in \mathcal{D}\backslash\{i\}}$. Therefore, we have the following theorem.

Definition 4.1. Denote the set of all possible values of Q_i as $\mathcal{S}_{Q_i}$, the set of all possible values of $\vec{Q}$ as $\mathcal{S}_Q := \prod_{i=1}^{D} \mathcal{S}_{Q_i}$, which is the Cartesian product of $\mathcal{S}_{Q_i}$, $i \in \mathcal{D}$, and the set of all possible values of $\{Q_j\}_{j \in \mathcal{D}\backslash\{i\}}$ as $\mathcal{S}_Q^{\bar{i}} := \prod_{j \in \mathcal{D}\backslash\{i\}} \mathcal{S}_{Q_j}$. Partition $\mathcal{S}_Q^{\bar{i}}$ into 2^{D-1} non-overlapping regions $\mathcal{S}_{\theta_v}^{\bar{i}}$, $v = 1, \ldots, 2^{D-1}$, such that the subset of queues with non-zero queue length is identical within each region, i.e., if $\{Q_j\}_{j \in \mathcal{D}\backslash\{i\}} \in \mathcal{S}_{\theta_v}^{\bar{i}}$, then $\{\Theta_j\}_{j \in \mathcal{D}\backslash\{i\}} = \theta_v$.

Theorem 4.1. $\forall \vec{Q}_t = \vec{k} \in \mathcal{S}_{Q_i} \times \mathcal{S}_{\theta_v}^{\bar{i}}$ and $\forall \vec{Q}_{t+1} = \vec{h} \in \mathcal{S}_{Q_i} \times \mathcal{S}_{\theta_w}^{\bar{i}}$, the values of $p_{(l_i,\vec{k},\vec{h})}^{n_i}$ are the same and can be denoted as $p_{(l_i,\theta_v,\theta_w)}^{n_i}$.

The proof of Theorem 4.1 is straightforward from (4.2). Therefore, we try to derive the value of $p_{(l_i,\theta_v,\theta_w)}^{n_i}$, which equals

$$p_{(l_i,\theta_v,\theta_w)}^{n_i} = \frac{p_{(\theta_v,\theta_w)}^{(l_i,n_i)}}{p_{\theta_v}^{l_i}}, \tag{4.6}$$

where

$$p^{(l_i,n_i)}_{(\theta_v,\theta_w)} = \text{Pr.}\{H_{i,t} = l_i, H_{i,t+1} = n_i | \{\Theta_{j,t}\}_{j\in\mathcal{D}\setminus\{i\}} = \theta_v, \{\Theta_{j,t+1}\}_{j\in\mathcal{D}\setminus\{i\}} = \theta_w\}.$$

(4.7)

$$p^{l_i}_{\theta_v} = \text{Pr.}\{H_{i,t} = l_i | \{\Theta_{j,t}\}_{j\in\mathcal{D}\setminus\{i\}} = \theta_v\}.$$

(4.8)

1. Derivation of $p^{l_i}_{\theta_v}$

   Since $H_{i,t} = l_i$, we have $SINR_{i,t} \in [\chi_{(l_i-1)}, \chi_{l_i})$. Therefore, according to (4.2), $\vec{\gamma}_{i,t}$ at time slot t belongs to the convex polyhedron $\Upsilon_{l_i} := \{\vec{\gamma}_i | \gamma_{ii} - \chi_{(l_i-1)}\sum_{j\in\mathcal{D}\setminus\{i\}}\gamma_{ji}\Theta_{j,t} \geq \chi_{(l_i-1)}, \gamma_{ii} - \chi_{l_i}\sum_{j\in\mathcal{D}\setminus\{i\}}\gamma_{ji}\Theta_{j,t} < \chi_{l_i}, \vec{\gamma}_i \geq 0\}$. The (virtual) SNR regions corresponding to the channel state l_i and $l_i + 1$ are separated by the hyperplane $\gamma_{ii} - \chi_{l_i}\sum_{j=1, j\neq i}^{D}\gamma_{ji}\Theta_{j,t} = \chi_{l_i}$. Assume there are two links in the system, an illustration of the SINR regions and its equivalent (virtual) SNR vector regions of link 1 when the channel state $H_{1,t} = 1, 2, 3$ and $\{\Theta_{2,t}\} = \{1\}$ is shown in Fig. 4.3. Therefore, the steady-state probability that $H_{i,t} = l_i$ given that $\{\Theta_{j,t}\}_{j\in\mathcal{D}\setminus\{i\}} = \theta_v$ can be derived as

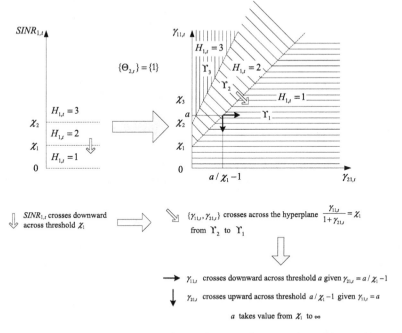

Fig. 4.3 (Virtual) SNR regions corresponding to the channel states and possible transitions of H_i from l_i into n_i when $v = w$

$$p_{\theta_v}^{l_i} = \int_{\Upsilon_{l_i}} f(\vec{\gamma}_i) d\vec{\gamma}_i = \int_{\Upsilon_{l_i}} \prod_{j \in \mathcal{D}} (f(\gamma_{ji}) d\gamma_{ji}). \qquad (4.9)$$

where $f(\vec{\gamma}_i)$ is the joint probability distribution function (pdf) of the stationary random process $\{\gamma_{ji,t}\}_{j \in \mathcal{D}}$, and $f(\gamma_{ji})$ is the pdf of $\gamma_{ji,t}$. The second equality is due to the independence between the r.v. elements in the set $\{\gamma_{ji}\}_{j \in \mathcal{D}}$. For a Rayleigh fading channel with additive white Gaussian noise, the received instantaneous SNR, γ_{ji}, is exponentially distributed with mean $\bar{\gamma}_{ji}$.

Therefore, $p_{\theta_v}^{l_i}$ can be derived by the integration of a multivariate exponential function over a convex polyhedron, which can be equivalently written as

$$p_{\theta_v}^{l_i} = \int \cdots \int_0^{\infty} \left(\int_{lb(l_i, \mathcal{J}_t)}^{ub(l_i, \mathcal{J}_t)} f(\gamma_{ii}) d\gamma_{ii} \right) \prod_{j \in \mathcal{J}_t} f(\gamma_{ji}) d\gamma_{ji}, \qquad (4.10)$$

where $ub(l_i, \mathcal{J}_t) = \chi_{l_i} + \chi_{l_i} \sum_{j \in \mathcal{J}_t} \gamma_{ji}$, $lb(l_i, \mathcal{J}_t) = \chi_{l_i-1} + \chi_{l_i-1} \sum_{j \in \mathcal{J}_t} \gamma_{ji}$. Therefore, the integration limits of γ_{ii} can be written as affine functions of $\gamma_{ji}, j \in \mathcal{J}_t$, while the integration limits of $\gamma_{ji}, j \in \mathcal{J}_t$ are all from 0 to ∞. Therefore, the closed-form expression for $p_{\theta_v}^{l_i}$ can be written as

$$p_{\theta_v}^{l_i} = \prod_{j \in \mathcal{J}_t} \frac{1}{\bar{\gamma}_{ji}} \left(\frac{\exp(-\chi_{(l_i-1)}/\bar{\gamma}_{ii})}{\prod_{k \in \mathcal{J}_t} (1/\bar{\gamma}_{ki} + \chi_{(l_i-1)}/\bar{\gamma}_{ii})} - \frac{\exp(-\chi_{l_i}/\bar{\gamma}_{ii})}{\prod_{k \in \mathcal{J}_t} (1/\bar{\gamma}_{ki} + \chi_{l_i}/\bar{\gamma}_{ii})} \right). \qquad (4.11)$$

2. Derivation of $p_{(\theta_v, \theta_w)}^{(l_i, n_i)}$

The transition of channel state H_i from l_i to n_i can be due to two categories of factors: (1) the fading of link i and its potential interfering links; (2) the variation of the set of interfering links of link i from θ_v to θ_w due to the changing backlog status of the other links. Therefore, we consider that $v = w$ and $v \neq w$, respectively, and derive the value of $p_{(\theta_v, \theta_w)}^{(l_i, n_i)}$ under each scenario. In the former scenario, factor (2) doesn't exist and we can focus on the SINR variation due to the fading effects.

a) When $v = w$: the values of $\{\Theta_{j,t}\}_{j \in \mathcal{D} \setminus \{i\}}$ and $\{\Theta_{j,t+1}\}_{j \in \mathcal{D} \setminus \{i\}}$ remain the same during two consecutive time slots, and thus the set of interference links $\{I_{ji}\}_{j \in \mathcal{J}_t}$ and $\{I_{ji}\}_{j \in \mathcal{J}_{t+1}}$ remain the same in time slot t and $t + 1$. Therefore, the transition of $H_{i,t}$ in state l_i to $H_{i,t+1}$ in state n_i can only be due to the variations of the (virtual) SNR vector $\{\gamma_{ji}\}_{j \in \{i\} \cup \mathcal{J}_t}$ in time slot t and $t + 1$. In [20], in order to derive the transition probability of the FSMC based on SNR partition, the product of the SNR level crossing rate and the time slot interval is used to approximate the joint probability that the channel states are in adjacent states in time slot t and $t + 1$, respectively. Furthermore, it assumes that the joint probabilities that the channel states are in different and non-adjacent states in two consecutive time slots are zero. In this chapter, we use a similar method to approximate the transition probabilities of the FSMC based on SINR partition. However, since the state transition of the SINR-based FSMC can be due to the variations of any of the (virtual) SNR elements in the (virtual) SNR vector $\{\gamma_{ji}\}_{j\in\{i\}\cup\mathcal{J}_t}$, the derivation method is much more complicated. We first use a similar assumption as in the SNR-based FSMC that the state transition of $H_{i,t}$ can only occur between adjacent states, i.e., $p_{(\theta_v,\theta_w)}^{(l_i,n_i)} = 0$, if $|n_i - l_i| > 1$. Then, we try to derive the transition probabilities between adjacent states as

$$p_{(\theta_v,\theta_w)}^{(l_i,l_i+1)} \approx NI(\chi_{l_i})\Delta T, \ p_{(\theta_v,\theta_w)}^{(l_i,l_i-1)} \approx NI(\chi_{(l_i-1)})\Delta T, \tag{4.12}$$

where $NI(\chi_{l_i})$ is "level crossing rate" in terms of the SINR, i.e., the expected number of times per second the SINR passes downward across the threshold χ_{l_i}. This approximation is similar to the method in [20]. $NI(\chi_{l_i})\Delta T$, whose value is smaller than one, can be explained as the probability that the SINR passes downward across the threshold χ_{l_i} in a time slot interval ΔT.

In order to derive the value of $NI(\chi_{l_i})\Delta T$, we consider a small time interval $\Delta t \rightarrow 0$. Therefore, $NI(\chi_{l_i})\Delta t$ can be explained as the probability that the SINR passes downward across the threshold χ_{l_i} in a small interval Δt.

Theorem 4.2. *We find the SINR value of link i crosses downward the threshold χ_{l_i} in a small time interval Δt if one of the mutually exclusive and exhaustive eventualities in the set $\{E_j\}_{j\in\{i\}\cup\mathcal{J}_t}$ occurred, where E_j is defined as the event that the (virtual) SNR γ_{ji} passes downward (or upward) across a threshold Γ_{j,l_i}, which equals*

$$\Gamma_{j,l_i} = \begin{cases} \frac{1}{\chi_{l_i}}(\gamma_{ii} - \chi_{l_i}(1 + \sum_{k\in\mathcal{D}\setminus\{i,j\}} \gamma_{ki}\Theta_{j,t})), & \text{if } j \neq i, \\ \chi_{l_i}(1 + \sum_{k\in\mathcal{D}\setminus\{i\}} \gamma_{ki}\Theta_{j,t}), & \text{if } j = i, \end{cases} \tag{4.13}$$

while the other (virtual) SNR values in the set $\{\gamma_{ki}\}_{k\in\{i\}\cup\mathcal{J}_t}$ remain unchanged. The probability of event E_j can be calculated as

$$\Pr.(E_j) = \int \cdots \int N_j(\Gamma_{j,l_i})\Delta t \prod_{k\in\{i\}\cup\mathcal{J}_t\setminus\{j\}} f(\gamma_{ki})\mathrm{d}\gamma_{ki}, \tag{4.14}$$

where $N_j(\Gamma)$, $j = 1,\ldots,D$ is the level crossing rate of the (virtual) SNR γ_{ji} at Γ, which is the expected number of times per second the (virtual) SNR γ_{ji} passes downward (or upward) across the threshold Γ, and can be calculated according to the Doppler shift f_m and the normalized threshold $\Gamma/\bar{\gamma}_{ji}$ [20].

Proof. Due to the equivalence between the SINR region $[\chi_{(l_i-1)}, \chi_{l_i})$ and the (virtual) SNR region Υ_{l_i} when the channel state $H_{i,t} = l_i$, $(l_i = 1,\ldots,L)$, the SINR value of link i crosses downward the threshold χ_{l_i} if and only if the (virtual) SNR vector $\vec{\gamma}_i$ passes across the hyperplane

$\gamma_{ii} - \chi_{l_i} \sum_{j \in \mathcal{D} \setminus \{i\}} \gamma_{ji} \Theta_{j,t} = \chi_{l_i}$ from the convex polyhedron Υ_{l_i+1} to Υ_{l_i} in a small interval Δt. Consider a (virtual) SNR γ_{ji} for $j \in \{i\} \bigcup \mathcal{J}_t$, given the values of the rest of the (virtual) SNR elements in the set $\{\gamma_{ki}\}_{k \in \{i\} \bigcup \mathcal{J}_t}$, we can derive the (virtual) SNR threshold Γ_{j,l_i} according to (4.13) so that if γ_{ji} crosses downward (or upward) Γ_{j,l_i} then the (virtual) SNR vector $\vec{\gamma}_i$ crosses the hyperplane $\gamma_{ii} - \chi_{l_i} \sum_{j=1, j \neq i}^{D} \gamma_{ji} \Theta_{j,t} = \chi_{l_i}$ from the convex polyhedron Υ_{l_i+1} to Υ_{l_i}. Since $N_j(\Gamma)\Delta t$ is the probability that the (virtual) SNR γ_{ji} passes downward (or upward) across the threshold Γ in a small interval Δt, and the values of the rest of the (virtual) SNR elements in the set $\{\gamma_{ki}\}_{k \in \{i\} \bigcup \mathcal{J}_t}$ can be taken over the field of non-negative real numbers as long as $\Gamma_{j,l_i} > 0$ with the pdf function $f(\gamma_{ki})$, the probability of event E_j can be derived as (4.14).

Note that the probability of multiple simultaneous variations of the (virtual) SNR elements in the vector $\vec{\gamma}_i$ in a small interval Δt are prohibited in the sense that each such multiple event is of order $o(\Delta t)$. Therefore, we will find the SINR value of link i crosses downward the threshold χ_{l_i}, or equivalently the (virtual) SNR vector $\vec{\gamma}_i$ crosses the hyperplane $\gamma_{ii} - \chi_{l_i} \sum_{j \in \mathcal{D} \setminus \{i\}} \gamma_{ji} \Theta_{j,t} = \chi_{l_i}$ from the convex polyhedron Υ_{l_i+1} to Υ_{l_i} in a time interval Δt, if one of the mutually exclusive and exhaustive eventualities E_j, $j \in \{i\} \bigcup \mathcal{J}_t$ occurred.

Example 4.1. As illustrated in Fig. 4.3 where there are two links and three channel states of $H_{i,t}$, $i = 1, 2$, we consider the link 1 and assume that $\{\Theta_{2,t}\} = \{\Theta_{2,t+1}\} = \theta_v = \{1\}$, i.e., link 1 suffers interference from link 2 in both time slots t and $t + 1$. $SINR_{i,t}$ crosses downward across threshold χ_1 so that $H_{1,t} = 2$ and $H_{1,t+1} = 1$. This is equivalent to $\{\gamma_{11,t}, \gamma_{21,t}\}$ crosses across the hyperplane $\frac{\gamma_{11,t}}{1+\gamma_{21,t}} = \chi_1$ from Υ_2 to Υ_1, which can only happen when $\gamma_{11,t}$ crosses downward across threshold a while $\gamma_{21,t} = a/\chi_1 - 1$ remains unchanged or $\gamma_{21,t}$ crosses upward across threshold $\gamma_{21,t} = a/\chi_1 - 1$ while $\gamma_{11,t} = a$ remains unchanged, where a takes values over the region $[\chi_1, \infty)$.

According to Theorem 4.2, we have

$$NI(\chi_{l_i})\Delta t = \sum_{j \in \mathcal{D}} \Pr.(E_j). \tag{4.15}$$

Let both sides of (4.15) be divided by Δt, we can derive the level crossing rate $NI(\chi_{l_i})$ of the SINR value of link i. Combining (4.15) with (4.12), we have

$$p_{(\theta_v, \theta_v)}^{(l_i, l_i+1)} \approx \sum_{j=1}^{D} \int \cdots \int_0^{\infty} N_j(\Gamma_{j,l_i})\Delta T \prod_{k \in \{i\} \bigcup \mathcal{J}_t \setminus \{j\}} f(\gamma_{ki})d\gamma_{ki}, \; p_{(\theta_v, \theta_v)}^{(l_i, l_i-1)}$$

$$\approx \sum_{j=1}^{D} \int \cdots \int_0^{\infty} N_j(\Gamma_{j,(l_i-1)})\Delta T \prod_{k \in \{i\} \bigcup \mathcal{J}_t \setminus \{j\}} f(\gamma_{ki})d\gamma_{ki}. $$

$$\tag{4.16}$$

b) When $v \neq w$: the values of $\{\Theta_{j,t}\}_{j \in \mathcal{D}\setminus\{i\}}$ and $\{\Theta_{j,t+1}\}_{j \in \mathcal{D}\setminus\{i\}}$ are different during two consecutive time slots. Therefore, the transition of $H_{i,t}$ from state l_i to $H_{i,t+1}$ in state n_i can be due to not only the variations of the (virtual) SNR vector $\{\gamma_{ji}\}_{j \in \{i\} \cup \mathcal{J}_t}$ in time slot t and $t+1$, but also the change of interference link set from $\{I_{ji}\}_{j \in \mathcal{J}_t}$ to $\{I_{ji}\}_{j \in \mathcal{J}_{t+1}}$. Therefore, we can no longer assume that the state transition of $H_{i,t}$ can only occur between adjacent states. Given θ_v and θ_w, we find the channel state $H_{i,t}$ in l_i and $H_{i,t+1}$ in n_i, respectively, if one of the three following (mutually exclusive and exhaustive) eventualities occurred:

1) that due to (virtual) SNR variations from $\bar{\gamma}_{i,t}$ to $\bar{\gamma}_{i,t+1}$, we would have found the channel state $H_{i,t}$ in state l_i and $\hat{H}_{i,t+1}$ in state $l_i + 1$ at the beginning of time slot t and $t+1$, respectively, if the value of $\{\hat{\Theta}_{j,t+1}\}_{j \in \mathcal{D}\setminus\{i\}}$ remains unchanged as $\{\Theta_{j,t}\}_{j \in \mathcal{D}\setminus\{i\}}$, which equals θ_v; however, since $\{\Theta_{j,t+1}\}_{j \in \mathcal{D}\setminus\{i\}}$ changes to θ_w at the beginning of time slot $t+1$, $H_{i,t+1}$ is in state n_i instead of state $l_i + 1$ with the same set of underlying (virtual) SNR vector values $\bar{\gamma}_{i,t+1}$; the multi-transition path described above is denoted as $H_{i,t} = l_i \rightarrow \hat{H}_{i,t+1} = l_i + 1 \rightarrow H_{i,t+1} = n_i$;
2) that similar to event 1), except that $\hat{H}_{i,t+1} = l_i - 1$, i.e., $H_i(t) = l_i \rightarrow \hat{H}_{i,t+1} = l_i - 1 \rightarrow H_{i,t+1} = n_i$;
3) that similar to event 1), except that $\hat{H}_{i,t+1} = l_i$, i.e., $H_i(t) = l_i \rightarrow \hat{H}_{i,t+1} = l_i \rightarrow H_{i,t+1} = n_i$.

Example 4.2. As illustrated in Fig. 4.4 where there are two links and three channel states of $H_{i,t}$, $i = 1, 2$, we consider the link 1 and assume that $\{\Theta_{2,t}\} = \theta_v = \{0\}$ and $\{\Theta_{2,t+1}\} = \theta_w = \{1\}$, i.e., link 1 suffers no interference from link 2 in time slot t, but it receives interference in time slot $t + 1$. We consider that $H_{1,t}$ is in state 3 and $H_{1,t+1}$ is in state 1 as shown in the upper part of Fig. 4.4 and examine the events that can cause this to happen as shown in the lower part of Fig. 4.4, which maps the SINR regions corresponding to the three states of $H_{1,t}$ to the (virtual) SNR regions. The solid lines mean that there is a change in the (virtual) SNR or the queue status, while the dotted lines mean no change has happened. First, assume that $\{\hat{\Theta}_{2,t+1}\}$ is unchanged as $\{\Theta_{2,t}\} = \{0\}$, and $\hat{H}_1(t + 1)$ can transit to the adjacent state 2 or remain in the same state 3 as $H_1(t)$. The (virtual) SNR region $\bar{\gamma}_{1,t+1}$ corresponding to both states are shown in Fig. 4.4 as light gray and medium gray areas. Second, since $\{\Theta_{2,t+1}\}$ is $\{1\}$ instead of $\{0\}$, $H_{1,t+1}$ is in state 1 instead of state 2 or state 3, and its corresponding (virtual) SNR region $\bar{\gamma}_{1,t+1}$ falls in the region surrounded by the bold lines, i.e., the overlapping regions of the horizontal stripped area with the light gray and medium gray areas. The multi-transition paths of the above two events are $H_{1,t} = 3 \rightarrow \hat{H}_{1,t+1} = 3 \rightarrow H_{1,t+1} = 1$ and $H_{1,t} = 3 \rightarrow \hat{H}_{1,t+1} = 2 \rightarrow H_{1,t+1} = 1$.

Therefore, we have

$$p_{(\theta_v,\theta_w)}^{(l_i,n_i)} = p_{(\theta_v,\theta_v)}^{(l_i,l_i+1)} \times \hat{p}_{(l_i+1,\theta_v,\theta_w)}^{n_i} + p_{(\theta_v,\theta_v)}^{(l_i,l_i-1)} \times \hat{p}_{(l_i-1,\theta_v,\theta_w)}^{n_i} + p_{(\theta_v,\theta_v)}^{(l_i,l_i)} \times \hat{p}_{(l_i,\theta_v,\theta_w)}^{n_i},$$

(4.17)

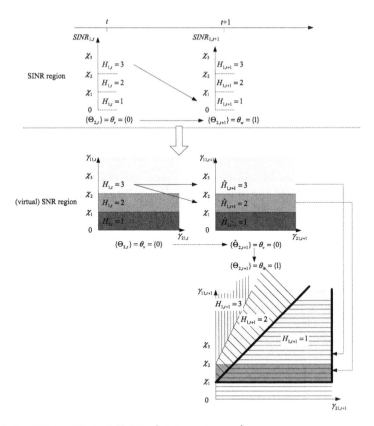

Fig. 4.4 Possible transitions of H_i from l_i into n_i when $v \neq w$

where $p_{(\theta_v,\theta_v)}^{(l_i,l_i)}$ (resp. $p_{(\theta_v,\theta_v)}^{(l_i,l_i+1)}$, $p_{(\theta_v,\theta_v)}^{(l_i,l_i-1)}$) is the conditional joint probability that $H_{i,t}$ is in state l_i and $\hat{H}_{i,t+1}$ is in state l_i (resp. $l_i + 1$, $l_i - 1$), given that $\{\hat{\Theta}_{j,t+1}\}_{j \in \mathcal{D} \setminus \{i\}} = \{\Theta_{j,t}\}_{j \in \mathcal{D} \setminus \{i\}} = \theta_v$, whose value can be derived from (4.16). On the other hand, $\hat{p}_{(l_i,\theta_v,\theta_w)}^{n_i}$ (resp. $\hat{p}_{(l_i+1,\theta_v,\theta_w)}^{n_i}$, $\hat{p}_{(l_i-1,\theta_v,\theta_w)}^{n_i}$) represents the conditional probability that $H_{i,t+1}$ is in state n_i, given that $\{\hat{\Theta}_{j,t+1}\}_{j \in \mathcal{D} \setminus \{i\}} = \theta_v$, $\hat{H}_{i,t+1}$ is in state l_i (resp. $l_i + 1$, $l_i - 1$), and $\{\Theta_{j,t+1}\}_{j \in \mathcal{D} \setminus \{i\}} = \theta_w$, i.e.,

$$\hat{p}_{(l_i,\theta_v,\theta_w)}^{n_i} = \mathrm{Pr}.\{H_i(t+1) = n_i | \hat{H}_i(t+1) = l_i, \{\hat{\Theta}_{j,t+1}\}_{j \in \mathcal{D} \setminus \{i\}}$$
$$= \theta_v, \{\Theta_{j,t+1}\}_{j \in \mathcal{D} \setminus \{i\}} = \theta_w\}. \qquad (4.18)$$

Given that $\hat{H}_i(t+1) = l_i$ and $\{\hat{\Theta}_{j,t+1}\}_{j \in \mathcal{D} \setminus \{i\}} = \theta_v$, the (virtual) SNR vector $\vec{\gamma}_{i,t+1}$ belongs to the convex polyhedron $\Upsilon_{l_i} := [\vec{\gamma}_i | \gamma_{ii} - \chi_{(l_i-1)} \sum_{j \in \mathcal{D} \setminus \{i\}} \gamma_{ji} \hat{\Theta}_{j,t+1} \geq \chi_{(l_i-1)} \cdot \gamma_{ii} - \chi_{l_i} \sum_{j \in \mathcal{D} \setminus \{i\}} \gamma_{ji} \hat{\Theta}_{j,t+1} < \chi_{l_i}, \vec{\gamma}_i \geq 0]$. Similarly, given $H_i(t+1) = n_i$ and $\{\Theta_{j,t+1}\}_{j \in \mathcal{D} \setminus \{i\}} =$

θ_w, $\vec{\gamma}_{i,t+1}$ also belongs to the convex polyhedron $\Upsilon_{n_i} := \{\vec{\gamma}_i | \gamma_{ii} - \chi_{(n_i-1)} \sum_{j \in \mathcal{D} \setminus \{i\}} \gamma_{ji} \Theta_{j,t+1} \geq \chi_{(n_i-1)}, \gamma_{ii} - \chi_{n_i} \sum_{j \in \mathcal{D} \setminus \{i\}} \gamma_{ji} \Theta_{j,t+1} < \chi_{n_i}, \vec{\gamma}_i \geq 0\}$. Therefore, the (virtual) SNR vector $\vec{\gamma}_{i,t+1}$ should belong to the convex polyhedron $\Upsilon_{l_i} \cap \Upsilon_{n_i}$, and

$$\hat{p}^{n_i}_{(l_i, \theta_v, \theta_w)} = \frac{\hat{p}^{l_i, n_i}_{(\theta_v, \theta_w)}}{p^{l_i+1}_{\theta_v}}, \tag{4.19}$$

where

$$\hat{p}^{(l_i, n_i)}_{(\theta_v, \theta_w)} = \Pr.\{\hat{H}_i(t+1) = l_i, H_i(t+1) = n_i | \{\hat{\Theta}_{j,t+1}\}_{j \in \mathcal{D} \setminus \{i\}} = \theta_v, \{\Theta_{j,t+1}\}_{j \in \mathcal{D} \setminus \{i\}} = \theta_w\}$$

$$= \int_{\Upsilon_{l_i} \cap \Upsilon_{n_i}} f(\vec{\gamma}_{i,t+1}) d\vec{\gamma}_{i,t+1}, \tag{4.20}$$

$$p^{l_i+1}_{\theta_v} = \Pr.\{\hat{H}_i(t+1) = l_i | \{\hat{\Theta}_{j,t+1}\}_{j \in \mathcal{D} \setminus \{i\}} = \theta_v\}. \tag{4.21}$$

The denominator of (4.19) can be derived according to (4.11). Similar to (4.10), the numerator of (4.19) is also the integration of a multivariate exponential over a convex polyhedron according to (4.20). However, since

$$\gamma_{ii} \in \left[\max\{lb(l_i, \mathcal{J}_{i,t}), lb(n_i, \mathcal{J}_{i,t+1})\}, \min\{ub(l_i, \mathcal{J}_{i,t}), ub(n_i, \mathcal{J}_{i,t+1})\} \right], \tag{4.22}$$

the integration limits of γ_{ii} cannot be written as affine functions of $\gamma_{ji}, j \in \mathcal{D} \setminus \{i\}$. Since integration over an arbitrary convex polyhedron is a non-trivial problem, and it has been shown that computing the volume of polytopes of varying dimension is NP-hard, we present a relatively simple method to calculate the integration of (4.20) in the Appendix 1.

Similar to (4.19), we can derive the values of $p^{n_i}_{(l_i+1, \theta_v, \theta_w)}$ and $p^{n_i}_{(l_i-1, \theta_v, \theta_w)}$. Taking these values into (4.17), we can derive the value of $p^{(l_i, n_i)}_{(\theta_v, \theta_w)}$ when $v \neq w$.

Now we have derived both the values of the denominator (by (4.11)) and numerator (by (4.16) when $v = w$ and by (4.17) when $v \neq w$) in (4.6), we can finalize the calculation of $p^{n_i}_{(l_i, \theta_v, \theta_w)}$ and thus $p^{n_i}_{(l_i, \vec{k}, \vec{h})}$.

Finally, we have

$$p^{\vec{n}}_{(\vec{l}, \vec{k}, \vec{h})} = \prod_{i=1}^{D} p^{n_i}_{(l_i, \vec{k}, \vec{h})}. \tag{4.23}$$

4.3.3 Steady-State Probability of Markov Chain

Define the transition probability matrix $\mathbf{P} = [p_{(\vec{l},\vec{k})}^{(\vec{n},\vec{h})}]$ and the steady-state probability matrix $\boldsymbol{\pi} = [\pi_{\vec{l},\vec{k}}]$, where $\pi_{\vec{l},\vec{k}} \equiv \lim_{t\to\infty} \mathrm{Pr}.\{\vec{H}_t = \vec{l}, \vec{Q}_t = \vec{k}\}$. Each element of the transition probability matrix $\mathbf{P}$ can be derived from (4.3) combining with (4.5) and (4.23).

Theorem 4.3. *The stationary distribution of the Markov chain $(\vec{H}_t, \vec{Q}_t)$ exists; $\boldsymbol{\pi}$ is unique, and $\boldsymbol{\pi} > 0$.*

Theorem 4.3 is proved in Appendix 2. Then, the stationary distribution of the ergodic process $\{\vec{H}_t, \vec{Q}_t\}$ can be uniquely determined from the balance equations

$$\boldsymbol{\pi} = \boldsymbol{\pi}\mathbf{P}, \quad \boldsymbol{\pi}\mathbf{e} = 1, \tag{4.24}$$

where $\mathbf{e}$ is the unity vector of dimension $(L \times (K + 1))^D$ and $\boldsymbol{\pi}$ can be derived as the normalized left eigenvector of $\mathbf{P}$ corresponding to eigenvalue 1.

4.4 Model Decomposition and Performance Approximation Using DSPNs

4.4.1 The DSPN Model

The analytical method in the previous section for the multi-user system faces the challenge of the exponentially enlarged state space, which makes it unacceptable for a large number of links. Since directly solving the queuing model suffers the high computational complexity, in this section, we formulate the SPN model of the above queuing system and use the model decomposition and an iteration procedure of SPN to simplify the analysis.

The $(2 \times D)$-dimensional discrete-time Markov chain $\{(\vec{H}_t, \vec{Q}_t), t = 0, 1, \ldots\}$ can be seen as a sampled-time Markov chain of a continuous-time semi-Markov process sampled at every ΔT interval, while the continuous-time semi-Markov process can be modeled as a DSPN. The DSPN consists of a SPN for representing service processes and a DSPN for representing queuing processes. The SPN, as shown in Fig. 4.5a, is composed of D subnets and each subnet i corresponds to the L-state Markov modulated service process of link i. Each subnet is described by places $(\{h_{il}\}_{l=1}^L)$ and transitions $\{tr_i^{(l,n)}\}_{\substack{l,n=1 \\ l\neq n}}^L$. The DSPN, as shown in Fig. 4.5b, models the queuing behavior of the links and can be characterized by places $\{q_i\}_{i\in\mathcal{D}}$, and transitions $\{c_i\}_{i\in\mathcal{D}}$, $\{s_i\}_{i\in\mathcal{D}}$. The meanings of all the places and transitions are described as follows.

- h_{il}: a place for the l-th channel state of link i.

Fig. 4.5 The DSPN Model
of FR strategy. (**a**) The
service process in SPN, (**b**)
The queue process in DSPN

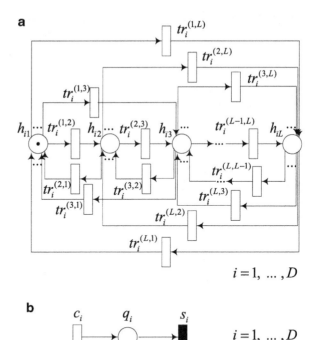

- $tr_i^{(l,n)}$: exponentially-distributed timed transitions for the channel state transitions of link i. When $tr_i^{(l,n)}$ fires, the channel state transits from l to n. The firing rate of $tr_i^{(l,n)}$ can be derived as $\rho_i^{(l,n)} = p_{(l,\vec{k},\vec{h})}^n/\Delta T$, where $p_{(l,\vec{k},\vec{h})}^n$ can be obtained by (4.6). Therefore, $\rho_i^{(l,n)}$ depends on the queue states of the other links before and after h_{il} transits to h_{in}, i.e., whether $M(q_j)$, $(j \in \mathcal{D}\backslash\{i\})$, is equal to or larger than zero, where $M(\cdot)$ is a mapping function from a place to the number of tokens assigned to it. Note that transitions from any channel state h_{il}, $l \in \{1,\ldots,L\}$ to any other channel states h_{in}, $n \in \{1,\ldots,L\}\backslash\{i\}$ are possible as proved in Lemma 4.1.
- q_i: a place for the queue state of link i.
- c_i: an exponentially-distributed timed transition denoting new packet arrivals from link i, with firing rate λ_i. When it fires, one packet arrives at the queue place q_i.
- s_i: a deterministic timed transitions for service process. When it fires, one packet is transmitted from the queue place q_i. Its firing rate μ_i depends on the marking of the places $\{h_{il}\}_{l=1}^L$, i.e.,

$$\mu_i = \lfloor \frac{R_l \Delta T}{B} \rfloor / \Delta T, \text{ if } M(h_{il}) = 1, \ l = 1,\ldots,L, \tag{4.25}$$

where $M(h_{il})$ is either 1 or 0, which represents whether link i is in its l-th channel state or not.

4.4.2 Model Decomposition and Iteration

According to Sect. 2.2, the original DSPN can be decomposed into a set of "near-independent" subnets. By decomposition, the original multiuser system is represented by D subsystems, each of which consists of one subnet in Fig. 4.5a and one subnet in Fig. 4.5b. Obviously, if each subsystem can be analyzed separately, the model decomposition can significantly reduce the size of the state space in the analysis and achieves better performance in computational complexity.

However, unfortunately, such model decomposition is not "clean", i.e., there exist interactions among subsystems, which is the same as those in Sect. 3.3.2. Specifically, for any subsystem $i \in \mathcal{D}$, the firing rate of transition s_i depends on the marking of the places $\{h_{il}\}_{l=1}^{L}$, which in turn depends on the markings of the queue places q_j of those links $j \in \mathcal{D}\backslash\{i\}$.

In order to solve this dilemma, we use the same methods have been proposed in Sect. 3.3.2. First, the steady-state probabilities of the markings instead of the instant markings of the other subsystems are used as the input of subsystem i in order to derive its steady-state probabilities of the markings. Next, fixed point iteration is used to deal with the cycles in the model solution process.

Let $\{H_{i,t}, Q_{i,t}\}$ denote the sampled-time Markov chain for the i-th DSPN subsystem. Let $\boldsymbol{\pi}_i := [\pi_{l_i,k_i}^{i}]$ denote the steady-state probabilities of $\{H_{i,t}, Q_{i,t}\}$, where $\pi_{l_i,k_i}^{i} \equiv \lim_{t\to\infty} \Pr.\{H_{i,t} = l_i, Q_{i,t} = k_i\}$. In order to derive the steady-state probabilities $\boldsymbol{\pi}_i$ of subsystem i, we have to first derive the transition probability matrix $\mathbf{P}_i = [p_{(l_i,k_i)}^{(n_i,h_i)}]$. First, $p_{(l_i,k_i)}^{h_i}$ can be derived according to (4.4). Next, we try to analyze the transition probability of the channel state from l_i to n_i given the queue states (k_i, h_i). According to (4.6), we can derive the value of $p_{(l_i,\vec{k},\vec{h})}^{n_i}$. However, since only the queue state (k_i, h_i) of subsystem i is given instead of $(\vec{k}, \vec{h})$, we assume that the steady state probabilities of all the other subsystems $\{\boldsymbol{\pi}_j\}_{j\in\mathcal{D}\backslash\{i\}}^{D}$ are known and derive the approximate transition probability $\tilde{p}_{(l_i,k_i,h_i)}^{n_i}$ as follows:

$$\tilde{p}_{(l_i,k_i,h_i)}^{n_i} = \sum_{\substack{\{k_j\}_{j\in\mathcal{D}\backslash\{i\}}\in\mathcal{S}_Q^{\vec{j}} \\ \{h_j\}_{j\in\mathcal{D}\backslash\{i\}}\in\mathcal{S}_Q^{\vec{j}}}} p_{(l_i,\vec{k},\vec{h})}^{n_i} \prod_{j\in\mathcal{D}\backslash\{i\}} \pi_{k_j,h_j}, \qquad (4.26)$$

where $\pi_{k_j,h_j} \equiv \lim_{t\to\infty} \Pr.\{Q_{j,t} = k_j, Q_{j,t+1} = h_j\}$ is the joint steady-state probability that the queue length of link j is k_j in time slot t and h_j in time slot $t+1$. Therefore, we have,

$$\pi_{k_j,h_j} = \sum_{l_j=1}^{L} p_{l_j,k_j}^{h_j} \pi_{l_j,k_j}^{j}, \qquad (4.27)$$

where $p_{l_j,k_j}^{h_j}$ can be obtained by (4.4).

Finally, we have

$$\tilde{p}_{(l_i,k_i)}^{(n_i,h_i)} = p_{(l_i,k_i)}^{h_i} \tilde{p}_{(l_i,k_i,h_i)}^{n_i},\tag{4.28}$$

which gives the transition probability matrix $\mathbf{P}_i$, and the steady state probabilities π_i can be derived similar to (4.24).

Define $\mathbf{x}_j := \{\pi_{k_j,h_j}\}_{k_j,h_j=0}^{K}$. According to (4.26), the solution π_i for the i-th subsystem can be obtained only when the measure $\{\mathbf{x}_j\}_{j \in \mathcal{D}\setminus\{i\}}$ are known so that the transition matrix $\mathbf{P}_i$ can be derived, and the value of $\{\mathbf{x}_j\}_{j \in \mathcal{D}\setminus\{i\}}$ depends on the solutions of all the other subsystems $\{\pi_j\}_{j \in \mathcal{D}\setminus\{i\}}$ according to (4.27). Obviously, if the D subsystems are solved sequentially by index, the above requirement cannot be satisfied since only $\{\mathbf{x}_j\}_{j=1}^{i-1}$ are known when solving the i-th subsystem. In the following, we can also use the fixed point iteration method analyzed in Sect. 3.3.2 to solve this problem.

We get the steady state probabilities π by the fixed point iteration, so the performance metrics such as the average queue length, the mean throughput, the average packet delay and the packet dropping probability can be derived as in [22].

- The average queue length of link i equals

$$\overline{Q}_i = \sum_{k_i=0}^{K} \sum_{l_i=1}^{L} \pi_{l_i,k_i}^i k_i,\tag{4.29}$$

where $\sum_{l_i=1}^{L} \pi_{l_i,k_i}^i$ is the probability that $Q_{i,t} = k_i$.
- The mean throughput of link i in terms of packets/s can be expressed as

$$\overline{T}_i = \sum_{l_i=1}^{L} \sum_{k_i=1}^{K} T_{l_i,k_i} \pi_{l_i,k_i}^i,\tag{4.30}$$

where

$$T_{l_i,k_i} = \begin{cases} \lfloor \frac{r_{i,t}\Delta T}{B} \rfloor / \Delta T & \text{if } k_i \geq \lfloor \frac{r_{i,t}\Delta T}{B} \rfloor, \\ \frac{k_i}{\Delta T} & \text{if } k_i < \lfloor \frac{r_{i,t}\Delta T}{B} \rfloor, \end{cases}\tag{4.31}$$

is the service rate of link i in terms of packets/s when $H_{i,t} = l_i$ and $Q_{i,t} = k_i$. It depends on the minimum value of the channel transmission capability and the amount of packets in the queue of link i. Note that the service rate is 0 when queue i is empty ($k_i = 0$). Therefore, $\overline{T}_i$ is the sum over the whole system state space of the product between the service rate of link i in state (l_i, k_i) and the probability that the system is in state (l_i, k_i).

- The average packet delay of link i can then be calculated according to Little's Law as

$$\overline{D}_i = \overline{Q}_i / \overline{T}_i, \tag{4.32}$$

which is the average amount of time between the arrival and departure of a packet for link i. Note that the mean throughput $\overline{T}_i$ equals the effective arrival rate of link i, which is the average rate at which the packets enter queue i.

- Let $B^i_{l_i,k_i}$ be the random variable which represents the number of dropped packets of link i when $H_{i,t} = l_i$ and $Q_{i,t} = k_i$. Since $K + b = A_{i,t} + \max[0, k - \lfloor \frac{r_{i,t}\Delta T}{B} \rfloor]$, where b is the number of packets dropped during the t-th slot,

$$\Pr.(B^i_{l_i,k_i} = b) = \Pr.(A_{i,t} = K + b - \max[0, k_i - \lfloor \frac{r_{i,t}\Delta T}{B} \rfloor]). \tag{4.33}$$

Then, the packet dropping probability p^i_{d} of link i can be estimated as

$$
\begin{aligned}
p^i_{\mathrm{d}} &= \frac{\text{Average \# of packets dropped in a time slot}}{\text{Average \# of packets arrived in a time slot}} \\
&= \frac{\sum_{l_i=1}^{L} \sum_{k_i=0}^{K} \sum_{b=0}^{\infty} b \Pr.(B_{l_i,k_i} = b)\pi^i_{l_i,k_i}}{\lambda_i \Delta T}.
\end{aligned} \tag{4.34}
$$

4.5   Numerical Results

In this section, we verify our analytical model under different interference conditions by tuning the length of the potential interfering links as shown in Fig. 4.6. We consider the path loss channel model $28 + 40 \log_{10} d$ [25] for all the D2D links and potential interfering links, where d is the distance between the transmitter and receiver in meter. We normalize the distance between a transmitter and a receiver with mean SNR equals $0\,\mathrm{dB}$ to be 1. The distance between a transmitter and a receiver of a link is denoted as α. Note that we do not require the length of the links to be the same in our analytical model, and this assumption in our network topology is only to facilitate us to focus on the variation of the potential interfering link length. Assume that the distances between the pair of transmitters (resp. receivers) of link i and $i + 1$ are $\beta + \Delta\beta(i - 1)$ $(i = 1, \ldots, D - 1)$. Therefore, the length of the potential interfering link I_{ji} of link i, where $j > i$ (resp. $j < i$), is $\sqrt{\alpha^2 + \left(\sum_{k=i}^{j-1}(\beta + \Delta\beta(k - 1))\right)^2}$ (resp. $\sqrt{\alpha^2 + \left(\sum_{k=j}^{i-1}(\beta + \Delta\beta(k - 1))\right)^2}$). In this way, we can increase (resp. decrease) the length of all the potential interfering links of link i by increase (resp. decrease) the value of β. We add $\Delta\beta(i - 1)$ to β in the distance between the transmitters of link i and $i + 1$ to ensure that the

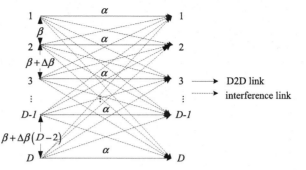

Fig. 4.6 The network topology. The distance between the transmitter and receiver of a link is denoted as α. The distance between the pair of transmitters (resp. receivers) of link i and $i + 1$ is $\beta + \Delta\beta(i - 1)$

Table 4.2 State space, iterations, and runtime with different number of D2D links

	State space			Runtime (s)	
D	Original	Decomposed	Iters	Numerics	Simulation
2	1.632e3	816	4	18	334 ($\lambda = 1$ k)
					970 ($\lambda = 100$ k)
3	1.332e6	816	4	25	657 ($\lambda = 1$ k)
					1,738 ($\lambda = 100$ k)
4	1.086e9	816	5	34	1136 ($\lambda = 1$ k)
					2,400 ($\lambda = 100$ k)

mean virtual SNR values $\bar{\gamma}_{ji}$ and $\bar{\gamma}_{ki}$ of any two potential interfering links I_{ji} and I_{ki} ($k \neq j \in \mathcal{D}\backslash\{i\}$) of link i are different, as required in (4.36).

The FSMC model has 16 states in total, and the SINR thresholds and the corresponding transmission rates in 1.4 MHz bandwidth for each service process are given in Table 4.1 as defined in the LTE system. The carrier frequency f and the time slot duration ΔT are set to 2 GHz and 1 ms, respectively. The velocity of the terminals is set to be 3 km/h so that the Doppler frequency becomes 5.56 Hz. We let the buffer size $K = 50$ packets, where the packet length $B = 50$ bits.

We numerically solve the decomposed Markov model using fixed point iteration and compare the performance measures with those obtained by discrete-event simulations of a D2D communications system with dynamic packet arrivals and full frequency reuse between D2D links. Both numeric method and simulation are implemented in Matlab and all experiments are run on 1.93 GHz PC with 1.87 GHz RAM. We increase the link number D from 2 to 4 and the state space of Markov models before and after decomposition are shown in Table 4.2. In the simulation, we generate Rayleigh fading channels by the Jakes Model using a U-shape Doppler power spectrum [26] for every pair of transmitter and receiver. In each simulated time slot, packets arrive to every queue according to Poisson distribution with mean $\lambda \Delta T$. For those D2D links with non-empty queues, we derive their respective SINR values in this time slot according to (4.2), where the channel gains of the D2D links and interfering links are generated by the Jakes Model. The corresponding transmission rates for the derived SINR values in this time slot can thus be derived

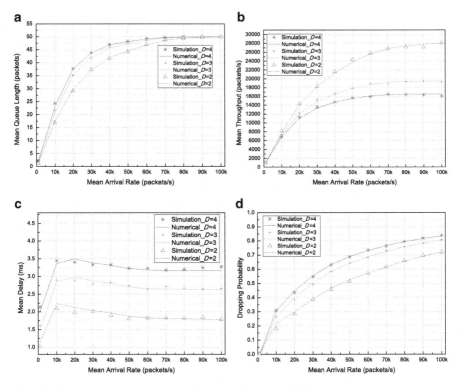

Fig. 4.7 Performance metrics versus packet arrival rate for D2D communications systems consisting of different numbers D of D2D links ($\alpha = 0.5$, $\beta = 0.3$, and $\Delta\beta = 0.01$). (**a**) Mean queue length, (**b**) mean throughput (packets/s), (**c**) mean delay (ms), (**d**) dropping probability

according to Table 4.1. The simulations are run over 10^5 time slots and the time-average performance measures of every D2D link are obtained. The simulation runtime is given in Table 4.2 which varies with both packet arrival rate λ and D2D link number D. In the numerical method, we set the initial steady-state probability vector $\boldsymbol{\pi}_i$ of every link i to be $\{\frac{1}{(K+1)\times L}, \ldots, \frac{1}{(K+1)\times L}\}$, and the initial vector of iteration variables $\{\mathbf{x}_1^0, \ldots, \mathbf{x}_D^0\}$ can be derived from (4.27). The number of iterations and runtime for convergence are given in Table 4.2. It can be observed that the runtime for the iterative numerics is much shorter than the simulation runtime.

Figure 4.7a–d show the mean queue length, mean throughput, mean delay and dropping probability averaged over all D2D links with varying arrival rates for D2D communications systems consisting of different numbers of links D, respectively. We choose $\alpha = 0.5$, $\beta = 0.3$ and $\Delta\beta = 0.01$. It can be seen that the numerical results match well with the simulation results under every configuration. As expected, the system performance in terms of all the above four measures degrade with the increasing number of links due to the growing amount of interference to every link. Figure 4.7a reveals that the mean queue length increases with the packet arrival rate and reaches the maximum buffer size 50 when the arrival rate reaches 100k packets/s. The variations of the mean throughput and dropping probability have a similar trend. Figure 4.7b shows that the difference in mean throughput when the numbers of links D are different grows larger with the increasing arrival rate. This is because the chances that the potential interfering links have data to transmit grow larger and thus the interference opportunities are increased for every link. The increase in mean throughput becomes insignificant when the arrival rate reaches 100k packets/s for every D. This is not surprising since it has almost reached the maximum transmission capacity of the system and the increasing arrival rate only results in increasing dropping probability. In Fig. 4.7d, it can be observed that the dropping probability is related to the maximum transmission capability of the system. Take $D = 4$ for example, the dropping probability reaches approximately 84 % when the arrival rate is 100k packets/s, which means the system is overload by a factor of 5.2 (i.e., dropping probability/(1-dropping probability)). On the other hand, Fig. 4.7b shows that the maximum transmission capability or the maximum mean throughput is 16k packets/s when $D = 4$, and 100k packets/s is indeed 5.2 times more than the maximum transmission capability of the system. Note that even in this saturated case, the amount of instantaneous interference received by a link can still be smaller than that under the infinite backlog traffic model, since the probability that the queue of any other link being empty cannot be zero as the Markov chain underlying the queuing system is irreducible. Figure 4.7c shows that the delay increases sharply when arrival rate increases from 1 to 20k packets/s, and then remains roughly the same when the arrival rate further increases. This is because when the arrival rate becomes larger than the maximum transmission capability of the system as discussed above, the system would have become overload if not for the packet dropping mechanism. Therefore, both the mean queue length and the mean throughput quickly reach their respective maximum values as revealed by Fig. 4.7a, b. By Little's Law, the mean delay also remains the same after that.

Figure 4.8a–d show the performance metrics with varying interference link length, where β ranges from 0.1 to 0.7 and the values of α and $\Delta\beta$ remain the same as above. The arrival rate is assumed to be 20K packets/s. The analytical results and the simulation results are very close, both of which improve with the increasing interference link length and decreasing interference from the other links. Note that when the interference link length is large, the performance gap between the scenarios with different numbers of links become small, since the difference in the amount of received interference is small in these topologies.

In the above numerical and simulation experiments, we assume that the packets arrive one at a time (as opposed to arriving in batches) following the Poisson process. However, we can extend the presented queuing model with the Batch Bernoulli arrival process by setting $Pr.(A_{i,t} = a)$ in (4.4) to be the probability mass function of a Binomial distribution. We illustrate the numerical and simulation results with respect to dropping probability for D2D communications systems consisting of different numbers D of D2D links under Batch Bernoulli arrival process in Fig. 4.9 with varying mean packet arrival rate and in Fig. 4.10 with varying interference link length. We observe that the numerical results match well

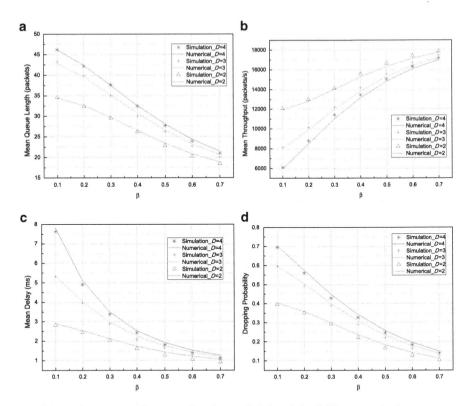

Fig. 4.8 Performance metrics versus interference link length for D2D communications systems consisting of different numbers D of D2D links ($\alpha = 0.5$, $\Delta\beta = 0.01$, mean arrival rate is 20K packets/s. Since the length of the potential interfering link I_{ji} of link i, where $j > i$ (resp. $j < i$) and $i, j \in \mathcal{D}$, is $\sqrt{\alpha^2 + \left(\sum_{k=i}^{j-1}(\beta + \Delta\beta(k-1))\right)^2}$ (resp. $\sqrt{\alpha^2 + \left(\sum_{k=j}^{i-1}(\beta + \Delta\beta(k-1))\right)^2}$), the interference link length varies with β on the x-axis). (**a**) Mean queue length, (**b**) mean throughput (packets/s), (**c**) mean delay (ms), (**d**) dropping probability

with the simulation results. Moreover, comparisons between Figs. 4.9 and 4.7d and between Figs. 4.10 and 4.8d show that the dropping probability under Poisson arrival process and Batch Bernoulli arrival process are quite similar.

4.6   Summary

In this chapter, we have developed a numerical method to investigate the performance of D2D communications with frequency reuse between D2D links and dynamic data arrival with finite-length queuing. The system behavior is formulated by a coupled processor queuing model, where the service process is characterized by a FSMC with each state corresponding to a certain SINR interval. We first construct

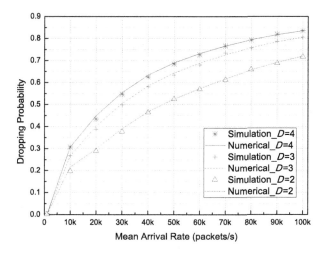

Fig. 4.9 Dropping probability versus packet arrival rate for D2D communications systems consisting of different numbers D of D2D links under Batch Bernoulli arrival process ($\alpha = 0.5$, $\beta = 0.3$, and $\Delta\beta = 0.01$)

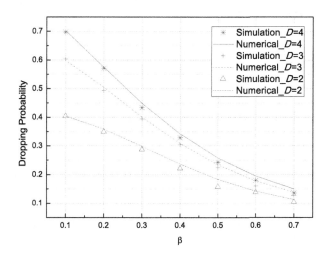

Fig. 4.10 Dropping probability versus interference link length for D2D communications systems consisting of different numbers D of D2D links under Batch Bernoulli arrival process ($\alpha = 0.5$, $\Delta\beta = 0.01$, mean arrival rate is 20K packets/s. Since the length of the potential interfering link I_{ji} of link i, where $j > i$ (resp. $j < i$) and $i, j \in \mathcal{D}$, is $\sqrt{\alpha^2 + \left(\sum_{k=i}^{j-1}(\beta + \Delta\beta(k-1))\right)^2}$ (resp. $\sqrt{\alpha^2 + \left(\sum_{k=j}^{i-1}(\beta + \Delta\beta(k-1))\right)^2}$), the interference link length varies with β on the x-axis)

the underlying DTMC of the queuing model and compute the state transition probabilities of the DTMC to derive its steady-state distribution. Since the state space of the DTMC grows exponentially with link number, we next formulate a DSPN model of the queueing system and use the model decomposition and iteration techniques in SPN to derive the approximate steady-sate distribution of the DTMC with low complexity. Finally, we obtain the performance metrics of the D2D communications from the steady-state distribution of the DTMC, whose accuracy has been verified by simulation results.

Appendix 1: Calculation of the Integration in (4.20)

Given $\{\hat{\Theta}_j\}_{j \in \mathcal{D}\setminus\{i\}} = \theta_v$, $\{\Theta_j\}_{j \in \mathcal{D}\setminus\{i\}} = \theta_w$ and $v \neq w$, we define $\mathcal{I}_i^v := \{I_{ji} | j \in \mathcal{D}\setminus\{i\} : \hat{\Theta}_j = 1\}$ as the subset of interfering links I_{ji} with $\hat{\Theta}_j = 1$, and $\mathcal{I}_i^w := \{I_{ji} | j \in \mathcal{D}\setminus\{i\} : \Theta_j = 1\}$ as the subset of interfering links I_{ji} with $\Theta_j = 1$. Furthermore, we define $\mathcal{I}_i^{vw} := \mathcal{I}_i^v \cap \mathcal{I}_i^w$, which could be an empty set $\emptyset$. Finally, we define $\mathcal{I}_i^{v\bar{w}} := \mathcal{I}_i^v \setminus \mathcal{I}_i^{vw}$ and $\mathcal{I}_i^{\bar{v}w} := \mathcal{I}_i^w \setminus \mathcal{I}_i^{vw}$, which are the difference of subsets $\mathcal{I}_i^v$ and $\mathcal{I}_i^w$ and the difference of subsets $\mathcal{I}_i^w$ and $\mathcal{I}_i^v$, respectively. Since $\mathcal{I}_i^v \neq \mathcal{I}_i^w$, the relationship between $\mathcal{I}_i^v$ and $\mathcal{I}_i^w$ can be divided into the following four mutually exclusive and exhaustive cases, as shown in Fig. 4.11:

(a) $\mathcal{I}_i^v = \emptyset$ or $\mathcal{I}_i^w = \emptyset$, and in both subcases, $\mathcal{I}_i^{vw} = \emptyset$; in the former subcase, we have $\mathcal{I}_i^{v\bar{w}} = \emptyset$, while in the latter subcase, we have $\mathcal{I}_i^{\bar{v}w} = \emptyset$; in the following three cases, we implicitly assume that $\mathcal{I}_i^v \neq \emptyset$ and $\mathcal{I}_i^w \neq \emptyset$;
(b) $\mathcal{I}_i^v \subset \mathcal{I}_i^w$ or $\mathcal{I}_i^w \subset \mathcal{I}_i^v$, and in the former subcase, we have $\mathcal{I}_i^{vw} = \mathcal{I}_i^v$, $\mathcal{I}_i^{v\bar{w}} = \emptyset$, while in the latter subcase, we have $\mathcal{I}_i^{vw} = \mathcal{I}_i^w$, $\mathcal{I}_i^{\bar{v}w} = \emptyset$;
(c) $\mathcal{I}_i^v \cap \mathcal{I}_i^w = \mathcal{I}_i^{vw} = \emptyset$, and in this case, we have $\mathcal{I}_i^{v\bar{w}} = \mathcal{I}_i^v$ and $\mathcal{I}_i^{\bar{v}w} = \mathcal{I}_i^w$;
(d) $\mathcal{I}_i^v \not\subseteq \mathcal{I}_i^w$, $\mathcal{I}_i^w \not\subseteq \mathcal{I}_i^v$, and $\mathcal{I}_i^{vw} \neq \emptyset$, and in this case, we have $\mathcal{I}_i^{v\bar{w}} \neq \emptyset$ and $\mathcal{I}_i^{\bar{v}w} \neq \emptyset$.

Denote the index set of $\mathcal{I}_i^{vw}$ as $\mathcal{J}^1$, i.e., $\mathcal{I}_i^{vw} = \{I_{ji}\}_{j \in \mathcal{J}^1}$. Similarly, denote the index sets of $\mathcal{I}_i^{v\bar{w}}$ and $\mathcal{I}_i^{\bar{v}w}$ as $\mathcal{J}^2$ and $\mathcal{J}^3$, respectively. In (4.19), the integration region of $\vec{\gamma}_i$ needs to obey the following constraints.

$$\chi_{(l_i-1)} \leq \frac{\gamma_{ii}}{1 + \sum_{j \in \mathcal{J}^1} \gamma_{ji} + \sum_{j \in \mathcal{J}^2} \gamma_{ji}} < \chi_{l_i}, \qquad (4.35)$$

$$\chi_{(n_i-1)} \leq \frac{\gamma_{ii}}{1 + \sum_{j \in \mathcal{J}^1} \gamma_{ji} + \sum_{j \in \mathcal{J}^3} \gamma_{ji}} < \chi_{n_i},$$

$$\vec{\gamma}_i \geq 0.$$

Let $\gamma s_1 := \sum_{j \in \mathcal{J}^1} \gamma_{ji}$, $\gamma s_2 := \sum_{j \in \mathcal{J}^2} \gamma_{ji}$, and $\gamma s_3 := \sum_{j \in \mathcal{J}^3} \gamma_{ji}$, which are all sum of independent exponential random variables. The pdf of γs_{id}, $id = 1, 2, 3$ has closed form expression as

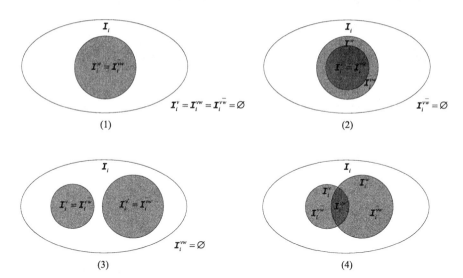

Fig. 4.11 The interfering link set

$$f_{\gamma_{S_{id}}}(x) = [\prod_{j \in \mathcal{J}^{id}} \frac{1}{\bar{\gamma}_{ji}}] \sum_{j \in \mathcal{J}^{id}} \frac{\exp(-x/\bar{\gamma}_{ji})}{\prod_{k \in \mathcal{J}^{id}\backslash\{j\}}(1/\bar{\gamma}_{ki} - 1/\bar{\gamma}_{ji})}, \qquad (4.36)$$

if $\{\gamma_{ji}\}_{j \in \mathcal{J}^{id}}$ have pairwise distinct mean $\bar{\gamma}_{ji}$.

Remark 4.1. Since the D2D terminals are randomly distributed in the cell in practical communications system, the probability that two D2D links have exactly the same mean SNR value is small. Therefore, the assumption that $\{\gamma_{ji}\}_{j \in \mathcal{J}^{id}}$ have pairwise distinct mean $\bar{\gamma}_{ji}$ is reasonable. Even if two D2D links do have the same mean SNR, we can add a very small number to one of the SNR to make the two values different, which will not have much impact on the performance evaluation results.

Now we try to calculate the integration in (4.20) under each of the four cases listed above.

1. *Case 1:* We consider that $\mathcal{I}_i^v = \emptyset$, while the subcase of $\mathcal{I}_i^w = \emptyset$ can be dealt with by a similar method. In this case, the polyhedron in (4.35) is equivalent to

$$\chi_{(l_i-1)} \leq \gamma_{ii} < \chi_{l_i}, \qquad (4.37)$$

$$\chi_{(n_i-1)} \leq \frac{\gamma_{ii}}{1 + \gamma_{S_3}} < \chi_{n_i},$$

$$\gamma_{ii} \geq 0, \ \gamma_{S_3} \geq 0.$$

which in turn is equivalent to

$$\chi_{(l_i-1)} \leq \gamma_{ii} < \chi_{l_i}, \tag{4.38}$$

$$\max\{0, \frac{\gamma_{ii}}{\chi_{n_i}} - 1\} \leq \gamma_{s_3} < \frac{\gamma_{ii}}{\chi_{(n_i-1)}} - 1.$$

2. *Case 2*: We consider that $\mathcal{I}_i^v \subset \mathcal{I}_i^w$, while the subcase of $\mathcal{I}_i^w \subset \mathcal{I}_i^v$ can be dealt with by a similar method. In this case, the polyhedron in (4.35) is equivalent to

$$\chi_{(l_i-1)} \leq \frac{\gamma_{ii}}{1 + \gamma_{s_1}} < \chi_{l_i}, \tag{4.39}$$

$$\chi_{(n_i-1)} \leq \frac{\gamma_{ii}}{1 + \gamma_{s_1} + \gamma_{s_3}} < \chi_{n_i},$$

$$\gamma_{ii} \geq 0, \ \gamma_{s_1} \geq 0, \ \gamma_{s_3} \geq 0.$$

which in turn is equivalent to

$$\chi_{(l_i-1)} \leq \gamma_{ii}, \tag{4.40}$$

$$\max\{0, \frac{\gamma_{ii}}{\chi_{l_i}} - 1\} \leq \gamma_{s_1} < \frac{\gamma_{ii}}{\chi_{(l_i-1)}} - 1,$$

$$\max\{0, \frac{\gamma_{ii}}{\chi_{n_i}} - \gamma_{s_1} - 1\} \leq \gamma_{s_3} < \frac{\gamma_{ii}}{\chi_{(n_i-1)}} - \gamma_{s_1} - 1.$$

3. *Case 3*: In this case, the polyhedron in (4.35) is equivalent to

$$\chi_{(l_i-1)} \leq \frac{\gamma_{ii}}{1 + \gamma_{s_2}} < \chi_{l_i}, \tag{4.41}$$

$$\chi_{(n_i-1)} \leq \frac{\gamma_{ii}}{1 + \gamma_{s_3}} < \chi_{n_i},$$

$$\gamma_{ii} \geq 0, \ \gamma_{s_2} \geq 0, \ \gamma_{s_3} \geq 0.$$

which in turn is equivalent to

$$\max\{\chi_{(l_i-1)}, \chi_{(n_i-1)}\} \leq \gamma_{ii}, \tag{4.42}$$

$$\max\{0, \frac{\gamma_{ii}}{\chi_{l_i}} - 1\} \leq \gamma_{s_2} < \frac{\gamma_{ii}}{\chi_{(l_i-1)}} - 1,$$

$$\max\{0, \frac{\gamma_{ii}}{\chi_{n_i}} - 1\} \leq \gamma_{s_3} < \frac{\gamma_{ii}}{\chi_{(n_i-1)}} - 1.$$

4. *Case 4*: In this case, the polyhedron in (4.35) is equivalent to

$$\chi_{(l_i-1)} \leq \frac{\gamma_{ii}}{1 + \gamma s_1 + \gamma s_2} < \chi_{l_i},$$

$$\chi_{(n_i-1)} \leq \frac{\gamma_{ii}}{1 + \gamma s_1 + \gamma s_3} < \chi_{n_i},$$

$$\gamma_{ii} \geq 0, \ \gamma s_1 \geq 0, \ \gamma s_2 \geq 0, \ \gamma s_3 \geq 0. \tag{4.43}$$

which in turn is equivalent to

$$\max\{\chi_{(l_i-1)}, \chi_{(n_i-1)}\} \leq \gamma_{ii},$$

$$\max\{0, \frac{\gamma_{ii}}{\chi_{l_i}} - \gamma s_1 - 1\} \leq \gamma s_2 < \frac{\gamma_{ii}}{\chi_{(l_i-1)}} - \gamma s_1 - 1,$$

$$\max\{0, \frac{\gamma_{ii}}{\chi_{n_i}} - \gamma s_1 - 1\} \leq \gamma s_3 < \frac{\gamma_{ii}}{\chi_{(n_i-1)}} - \gamma s_1 - 1,$$

$$0 \leq \gamma s_1 < \min\{\frac{\gamma_{ii}}{\chi_{(l_i-1)}} - 1, \frac{\gamma_{ii}}{\chi_{(n_i-1)}} - 1\}. \tag{4.44}$$

In all the above cases, we try to have the integration limits of γ_{ii}, γs_1, γs_2, and γs_3 as affine functions. Since Case 4 is the most complex one and the other cases can be considered as special circumstances of Case 4, we will only discuss the integration of (4.20) under Case 4 in details due to space limitation.

1) If $l_i = n_i$,

- and if $\chi_{(l_i-1)} \leq \gamma_{ii} < \chi_{l_i}$, we have $A1$ equals

$$\int_{\chi_{(l_i-1)}}^{\chi_{l_i}} f(\gamma_{ii})d\gamma_{ii} \int_0^{\frac{\gamma_{ii}}{\chi_{(l_i-1)}}-1} f_{\gamma s_1}(\gamma s_1)d\gamma s_1 \int_0^{\frac{\gamma_{ii}}{\chi_{(l_i-1)}}-\gamma s_1-1}$$

$$f_{\gamma s_2}(\gamma s_2)d\gamma s_2 \int_0^{\frac{\gamma_{ii}}{\chi_{(l_i-1)}}-\gamma s_1-1} f_{\gamma s_3}(\gamma s_3)d\gamma s_3, \tag{4.45}$$

- and if $\gamma_{ii} \geq \chi_{l_i}$, $\gamma s_1 \geq \frac{\gamma_{ii}}{\chi_{l_i}} - 1$, we have $A2$ equals

$$\int_{\chi_{l_i}}^{\infty} f(\gamma_{ii})d\gamma_{ii} \int_{\frac{\gamma_{ii}}{\chi_{l_i}}-1}^{\frac{\gamma_{ii}}{\chi_{(l_i-1)}}-1} f_{\gamma s_1}(\gamma s_1)d\gamma s_1 \int_0^{\frac{\gamma_{ii}}{(\chi_{l_i}-1)}-\gamma s_1-1}$$

$$f_{\gamma s_2}(\gamma s_2)d\gamma s_2 \int_0^{\frac{\gamma_{ii}}{(\chi_{l_i}-1)}-\gamma s_1-1} f_{\gamma s_3}(\gamma s_3)d\gamma s_3, \tag{4.46}$$

- else if $\gamma_{ii} \geq \chi_{l_i}$, $\gamma s_1 < \frac{\gamma_{ii}}{\chi_{l_i}} - 1$, we have $A3$ equals

$$\int_{\chi_{l_i}}^{\infty} f(\gamma_{ii})d\gamma_{ii} \int_{0}^{\frac{\gamma_{ii}}{\chi_{l_i}}-1} f_{\gamma s_1}(\gamma s_1)d\gamma s_1 \int_{\frac{\gamma_{ii}}{\chi_{l_i}}-\gamma s_1-1}^{\frac{\gamma_{ii}}{\chi_{l_i}}-\gamma s_1-1}$$

$$f_{\gamma s_2}(\gamma s_2)d\gamma s_2 \int_{\frac{\gamma_{ii}}{\chi_{l_i}}-\gamma s_1-1}^{\frac{\gamma_{ii}}{\chi_{l_i}}-\gamma s_1-1} f_{\gamma s_3}(\gamma s_3)d\gamma s_3. \tag{4.47}$$

2) If $l_i > n_i$,

- and if $\chi_{(l_i-1)} \leq \gamma_{ii} < \chi_{l_i}$, we have $A1$ equals

$$\int_{\chi_{(l_i-1)}}^{\chi_{l_i}} f(\gamma_{ii})d\gamma_{ii} \int_{0}^{\frac{\gamma_{ii}}{\chi_{(l_i-1)}}-1} f_{\gamma s_1}(\gamma s_1)d\gamma s_1 \int_{0}^{\frac{\gamma_{ii}}{\chi_{(l_i-1)}}-\gamma s_1-1}$$

$$f_{\gamma s_2}(\gamma s_2)d\gamma s_2 \int_{\frac{\gamma_{ii}}{\chi_{n_i}}-\gamma s_1-1}^{\frac{\gamma_{ii}}{\chi_{(n_i-1)}}-\gamma s_1-1} f_{\gamma s_3}(\gamma s_3)d\gamma s_3, \tag{4.48}$$

- and if $\gamma_{ii} \geq \chi_{l_i}$, $\gamma s_1 \geq \frac{\gamma_{ii}}{\chi_{l_i}} - 1$, we have $A2$ equals

$$\int_{\chi_{l_i}}^{\infty} f(\gamma_{ii})d\gamma_{ii} \int_{\frac{\gamma_{ii}}{\chi_{l_i}}-1}^{\frac{\gamma_{ii}}{\chi_{(l_i-1)}}-1} f_{\gamma s_1}(\gamma s_1)d\gamma s_1 \int_{0}^{\frac{\gamma_{ii}}{\chi_{(l_i-1)}}-\gamma s_1-1}$$

$$f_{\gamma s_2}(\gamma s_2)d\gamma s_2 \int_{\frac{\gamma_{ii}}{\chi_{n_i}}-\gamma s_1-1}^{\frac{\gamma_{ii}}{\chi_{(n_i-1)}}-\gamma s_1-1} f_{\gamma s_3}(\gamma s_3)d\gamma s_3, \tag{4.49}$$

- else if $\gamma_{ii} \geq \chi_{l_i}$, $\gamma s_1 < \frac{\gamma_{ii}}{\chi_{l_i}} - 1$, we have $A3$ equals

$$\int_{\chi_{l_i}}^{\infty} f(\gamma_{ii})d\gamma_{ii} \int_{0}^{\frac{\gamma_{ii}}{\chi_{l_i}}-1} f_{\gamma s_1}(\gamma s_1)d\gamma s_1 \int_{\frac{\gamma_{ii}}{\chi_{l_i}}-\gamma s_1-1}^{\frac{\gamma_{ii}}{\chi_{(l_i-1)}}-\gamma s_1-1}$$

$$f_{\gamma s_2}(\gamma s_2)d\gamma s_2 \int_{\frac{\gamma_{ii}}{\chi_{n_i}}-\gamma s_1-1}^{\frac{\gamma_{ii}}{\chi_{(n_i-1)}}-\gamma s_1-1} f_{\gamma s_3}(\gamma s_3)d\gamma s_3. \tag{4.50}$$

3) If $l_i < n_i$,

- and if $\chi_{(n_i-1)} \leq \gamma_{ii} < \chi_{n_i}$, we have $A1$ equals

$$\int_{\chi_{(n_i-1)}}^{\chi_{n_i}} f(\gamma_{ii})d\gamma_{ii} \int_{0}^{\frac{\gamma_{ii}}{\chi_{(n_i-1)}}-1} f_{\gamma s_1}(\gamma s_1)d\gamma s_1 \int_{\frac{\gamma_{ii}}{\chi_{l_i}}-\gamma s_1-1}^{\frac{\gamma_{ii}}{\chi_{(l_i-1)}}-\gamma s_1-1}$$

$$f_{\gamma s_2}(\gamma s_2)d\gamma s_2 \int_{0}^{\frac{\gamma_{ii}}{\chi_{(n_i-1)}}-\gamma s_1-1} f_{\gamma s_3}(\gamma s_3)d\gamma s_3, \tag{4.51}$$

- and if $\gamma_{ii} \geq \chi_{n_i}$, $\gamma s_1 \geq \frac{\gamma_{ii}}{\chi_{n_i}} - 1$, we have $A2$ equals

$$\int_{\chi_{n_i}}^{\infty} f(\gamma_{ii}) d\gamma_{ii} \int_{\frac{\gamma_{ii}}{\chi_{n_i}}-1}^{\frac{\gamma_{ii}}{\chi_{(n_i-1)}}-1} f_{\gamma s_1}(\gamma s_1) d\gamma s_1 \int_{\frac{\gamma_{ii}}{\chi_{l_i}}-\gamma s_1-1}^{\frac{\gamma_{ii}}{\chi_{(l_i-1)}}-\gamma s_1-1}$$

$$f_{\gamma s_2}(\gamma s_2) d\gamma s_2 \int_{0}^{\frac{\gamma_{ii}}{\chi_{(n_i-1)}}-\gamma s_1-1} f_{\gamma s_3}(\gamma s_3) d\gamma s_3, \tag{4.52}$$

- else if $\gamma_{ii} \geq \chi_{n_i}$, $\gamma s_1 < \frac{\gamma_{ii}}{\chi_{n_i}} - 1$, we have $A3$ equals

$$\int_{\chi_{n_i}}^{\infty} f(\gamma_{ii}) d\gamma_{ii} \int_{0}^{\frac{\gamma_{ii}}{\chi_{n_i}}-1} f_{\gamma s_1}(\gamma s_1) d\gamma s_1 \int_{\frac{\gamma_{ii}}{\chi_{l_i}}-\gamma s_1-1}^{\frac{\gamma_{ii}}{\chi_{(l_i-1)}}-\gamma s_1-1}$$

$$f_{\gamma s_2}(\gamma s_2) d\gamma s_2 \int_{\frac{\gamma_{ii}}{\chi_{n_i}}-\gamma s_1-1}^{\frac{\gamma_{ii}}{\chi_{(n_i-1)}}-\gamma s_1-1} f_{\gamma s_3}(\gamma s_3) d\gamma s_3. \tag{4.53}$$

Therefore, $\hat{p}_{(\theta_v,\theta_w)}^{l_i+1,n_i} = A1 + A2 + A3$ when $l_i = n_i$, $l_i > n_i$, and $l_i < n_i$, respectively. Combining (4.36) with the above integrations, and after mathematical manipulation, we have:

1) if $l_i = n_i$,

$$\hat{p}_{(\theta_v,\theta_w)}^{l_i,n_i} = \frac{1}{\bar{\gamma}_{ii}} [\prod_{j \in \mathcal{J}^1} \frac{1}{\bar{\gamma}_{ji}}][\prod_{j' \in \mathcal{J}^2} \frac{1}{\bar{\gamma}_{j'i}}][\prod_{j'' \in \mathcal{J}^3} \frac{1}{\bar{\gamma}_{j''i}}] \sum_{j \in \mathcal{J}^1} \sum_{j' \in \mathcal{J}^2} \sum_{j'' \in \mathcal{J}^3}$$

$$\frac{\bar{\gamma}_{j'i}\bar{\gamma}_{j''i}(F(\chi_{(l_i-1)}) - F(\chi_{l_i}))}{\prod_{k \in \mathcal{J}^1 \setminus \{j\}} [\frac{1}{\bar{\gamma}_{ki}} - \frac{1}{\bar{\gamma}_{ji}}] \prod_{k' \in \mathcal{J}^2 \setminus \{j'\}} [\frac{1}{\bar{\gamma}_{k'i}} - \frac{1}{\bar{\gamma}_{j'i}}]} \frac{1}{\prod_{k'' \in \mathcal{J}^3 \setminus \{j''\}} [\frac{1}{\bar{\gamma}_{k''i}} - \frac{1}{\bar{\gamma}_{j''i}}]}, \tag{4.54}$$

where

$$F(a) = \frac{\exp(\frac{-a}{\bar{\gamma}_{ii}})\bar{\gamma}_{ii}^4\bar{\gamma}_{ji}(2a\bar{\gamma}_{j'i}\bar{\gamma}_{j''i} + \bar{\gamma}_{ii}(\bar{\gamma}_{j'i} + \bar{\gamma}_{j''i}))}{(\bar{\gamma}_{ii} + a\bar{\gamma}_{ji})(\bar{\gamma}_{ii} + a\bar{\gamma}_{j'i})(\bar{\gamma}_{ii} + a\bar{\gamma}_{j''i})}$$

$$\times \frac{1}{(a\bar{\gamma}_{j'i}\bar{\gamma}_{j''i} + \bar{\gamma}_{ii}(\bar{\gamma}_{j'i} + \bar{\gamma}_{j''i}))}. \tag{4.55}$$

2) if $l_i > n_i$,

$$\hat{p}_{(\theta_v,\theta_w)}^{l_i,n_i} = \frac{1}{\bar{\gamma}_{ii}} [\prod_{j \in \mathcal{J}^1} \frac{1}{\bar{\gamma}_{ji}}][\prod_{j' \in \mathcal{J}^2} \frac{1}{\bar{\gamma}_{j'i}}][\prod_{j'' \in \mathcal{J}^3} \frac{1}{\bar{\gamma}_{j''i}}] \sum_{j \in \mathcal{J}^1} \sum_{j' \in \mathcal{J}^2} \sum_{j'' \in \mathcal{J}^3}$$

$$\frac{\bar{\gamma}_{j'i}\bar{\gamma}_{j''i}\exp(\frac{1}{\bar{\gamma}_{j''i}})}{\displaystyle\prod_{k\in\mathcal{J}^1\setminus\{j\}}[\frac{1}{\bar{\gamma}_{ki}}-\frac{1}{\bar{\gamma}_{ji}}]\prod_{k'\in\mathcal{J}^2\setminus\{j'\}}[\frac{1}{\bar{\gamma}_{k'i}}-\frac{1}{\bar{\gamma}_{j'i}}]\prod_{k''\in\mathcal{J}^3\setminus\{j''\}}[\frac{1}{\bar{\gamma}_{k''i}}-\frac{1}{\bar{\gamma}_{j''i}}]}\qquad\frac{1}{}$$

$$(F'(\chi_{n_i},\chi_{(l_i-1)})-F'(\chi_{(n_i-1)},\chi_{(l_i-1)})-F'(\chi_{n_i},\chi_{l_i})+F'(\chi_{(n_i-1)},\chi_{l_i})),$$

(4.56)

where

$$F'(a,b)=\frac{\exp(-b(\frac{1}{a\bar{\gamma}_{j''i}}+\frac{1}{\bar{\gamma}_{ii}}))a^3\bar{\gamma}_{ii}^3\bar{\gamma}_{ji}^3\bar{\gamma}_{j''i}^3}{(\bar{\gamma}_{ii}+a\bar{\gamma}_{j''i})(a\bar{\gamma}_{ii}\bar{\gamma}_{j''i}+b\bar{\gamma}_{j'i}(\bar{\gamma}_{ii}+a\bar{\gamma}_{j''i}))}$$

$$\times\frac{1}{(b\bar{\gamma}_{ii}\bar{\gamma}_{ji}+a(b\bar{\gamma}_{ii}\bar{\gamma}_{j''i}+\bar{\gamma}_{ii}(\bar{\gamma}_{j''i}-\bar{\gamma}_{ji})))}.$$

(4.57)

3) if $l_i < n_i$, $\hat{p}_{(\theta_v,\theta_w)}^{l_i,n_i}$ can be derived according to (4.56) except that $\chi_{(n_i-1)}$ (resp. χ_{n_i}) and $\chi_{(l_i-1)}$ (resp. χ_{l_i}) switch places with each other.

Although we will not discuss the integration of (4.20) in case 1, 2 and 3 in detail, we will prove the following Lemma, which will be used in Appendix 2 for the prove of Theorem 4.3.

Lemma 4.1. *In Case 1 and Case 2, $\hat{p}_{(\theta_v,\theta_w)}^{l_i,n_i} > 0$ for any $l_i,n_i \in \{1,\ldots,L\}$ satisfying $l_i \geq n_i$ (resp. $l_i \leq n_i$) when $\mathcal{I}_i^v \subset \mathcal{I}_i^w$ (resp. $\mathcal{I}_i^w \subset \mathcal{I}_i^v$); In Case 3 and Case 4, $\hat{p}_{(\theta_v,\theta_w)}^{l_i,n_i} > 0$ for any $l_i,n_i \in \{1,\ldots,L\}$.*

Proof. In order to prove $\hat{p}_{(\theta_v,\theta_w)}^{l_i,n_i} > 0$, we need to show that the integration region $\Upsilon_{l_i} \cap \Upsilon_{n_i}$ is non-empty according to (4.20).

- In case 1, since (4.38) defines the integration region when $\mathcal{I}_i^v \subset \mathcal{I}_i^w$, we need to verify that for any $l_i,n_i \in \{1,\ldots,L\}$ satisfying $l_i \geq n_i$, the upper limit of integration for γs_3 is not always smaller than its lower limits when $\gamma_{ii} \in [\chi_{(l_i-1)},\chi_{l_i})$. In (4.38), if $l_i \geq n_i$, we have $\gamma_{ii} \geq \chi_{(l_i-1)} \geq \chi_{(n_i-1)}$. Therefore, the upper limits of integration for γs_3, i.e., $\frac{\gamma_{ii}}{\chi_{(n_i-1)}} - 1$ is larger than zero when $\gamma_{ii} > \chi_{(n_i-1)}$, and thus larger than the lower limits of integration for γs_3, i.e., $\max\{0, \frac{\gamma_{ii}}{\chi_{n_i}} - 1\}$. The scenario when $\mathcal{I}_i^w \subset \mathcal{I}_i^v$ can be proved in a similar way.
- In case 2, since (4.40) defines the integration region when $\mathcal{I}_i^v \subset \mathcal{I}_i^w$, we need to verify that for any $l_i,n_i \in \{1,\ldots,L\}$ satisfying $l_i \geq n_i$, the upper limits of integration for γs_1 and γs_3 are not always smaller than their corresponding lower limits when $\gamma_{ii} \in [\chi_{(l_i-1)},\infty)$. In (4.40), the upper limit of γs_1, i.e., $\frac{\gamma_{ii}}{\chi_{(l_i-1)}} - 1$ is larger than zero when $\gamma_{ii} > \chi_{(l_i-1)}$, and thus larger than the lower limits of integration for γs_1, i.e., $\max\{0, \frac{\gamma_{ii}}{\chi_{l_i}} - 1\}$. Furthermore, since $l_i \geq n_i$ and $\gamma s_1 < \frac{\gamma_{ii}}{\chi_{(l_i-1)}} - 1$, the upper limits of integration for γs_3, i.e., $\frac{\gamma_{ii}}{\chi_{(n_i-1)}} - \gamma s_1 - 1$ is larger than

zero, and thus larger than the lower limits of integration for γs_3, i.e., $\max\{0, \frac{\gamma_{ii}}{\chi_{n_i}} - \gamma s_1 - 1\}$. The scenario when $\mathcal{I}_i^w \subset \mathcal{I}_i^v$ can be proved in a similar way.

- In case 3, since (4.42) defines the integration region, we need to verify that for any $l_i, n_i \in \{1, \ldots, L\}$, the upper limits of integration for γs_2 and γs_3 are not always smaller than their corresponding lower limits when $\gamma_{ii} \in [\max\{\chi_{(l_i-1)}, \chi_{(n_i-1)}\}, \infty)$. In (4.42), the upper limits of integration for γs_2 and γs_3, i.e. $\frac{\gamma_{ii}}{\chi_{(l_i-1)}} - 1$ and $\frac{\gamma_{ii}}{\chi_{(n_i-1)}} - 1$ are larger than zero when $\gamma_{ii} > \max\{\chi_{(l_i-1)}, \chi_{(n_i-1)}\}$, and thus larger than their corresponding lower limits of integration, i.e., $\max\{0, \frac{\gamma_{ii}}{\chi_{l_i}} - 1\}$ and $\max\{0, \frac{\gamma_{ii}}{\chi_{n_i}} - 1\}$.
- In case 4, since A_1, A_2 and A_3 are all larger than zero when $l_i > n_i$, $l_i < n_i$ and $l_i = n_i$, $\hat{p}_{(\theta_v,\theta_w)}^{l_i,n_i} = A_1 + A_2 + A_3 > 0$ for any $l_i, n_i \in \{1, \ldots, L\}$.

Appendix 2: Proof of Theorem 4.3

We first prove the following lemmas.

Lemma 4.2. *The Markov chain $(\vec{H}_t, \vec{Q}_t)$ is irreducible, if $K \leq R_L \Delta T$.*

Proof. We can prove Lemma 4.2 by showing that for each transition from state $(\vec{l}, \vec{k})$ to $(\vec{n}, \vec{h})$, there exists a multi-transition path with non-zero probability, which is denoted as $(\vec{l}, \vec{k}) \longrightarrow (\vec{n}, \vec{h})$. Now we shall verify the following cases:

1) $(\vec{l}, \vec{k}) \longrightarrow (\vec{l}^*, \vec{k})$, for any $\vec{l}^* = \{l_i^* \in \{1, \ldots, L\} : R_{l_i^*} \Delta T \geq k_i\}_{i \in \mathcal{D}}$.

- First, we will prove that $p_{(\vec{l},\vec{k},\vec{k})}^{\{l_i+1\}_{i \in \mathcal{D}}} > 0$, $p_{(\vec{l},\vec{k},\vec{k})}^{\{l_i-1\}_{i \in \mathcal{D}}} > 0$ and $p_{(\vec{l},\vec{k},\vec{k})}^{\vec{l}} > 0$. From (4.23), we only need to prove that $p_{(l_i,\vec{k},\vec{k})}^{(l_i+1)} > 0$, $p_{(l_i,\vec{k},\vec{k})}^{(l_i-1)} > 0$, and $p_{(l_i,\vec{k},\vec{k})}^{l_i} > 0$. According to Theorem 4.1, this is equivalent to proving that $p_{(l_i,\theta_v,\theta_v)}^{(l_i+1)} > 0$, $p_{(l_i,\theta_v,\theta_v)}^{(l_i-1)} > 0$, and $p_{(l_i,\theta_v,\theta_v)}^{l_i} > 0$, where $\vec{k} \in S_{Q_i} \times S_{\theta_v}^i$. This is true from (4.6), since $p_{\theta_v}^{l_i} > 0$, and we have $p_{(\theta_v,\theta_v)}^{(l_i,l_i+1)} > 0$, $p_{(\theta_v,\theta_v)}^{(l_i,l_i-1)} > 0$, and $p_{(\theta_v,\theta_v)}^{(l_i,l_i)} > 0$ based on (4.16).
- Then, we will prove that $p_{\vec{l},\vec{k}}^{\vec{k}} > 0$. Since $A_{i,t} = k_i - \max[0, k_i - R_{l_i} \Delta T] \geq 0$, we have $p_{l_i,k_i}^{k_i} > 0$ from (4.4) and $p_{\vec{l},\vec{k}}^{\vec{k}} > 0$ from (4.5).
- Therefore, we have $p_{(\vec{l},\vec{k})}^{(\{l_i+1\}_{i \in \mathcal{D}}, \vec{k})} > 0$, $p_{(\vec{l},\vec{k})}^{(\{l_i-1\}_{i \in \mathcal{D}}, \vec{k})} > 0$ and $p_{(\vec{l},\vec{k})}^{(\vec{l},\vec{k})} > 0$ from (4.3), and there exists a multi-transition path from $(\vec{l}, \vec{k})$ to $(\vec{l}^*, \vec{k})$ as $(\vec{l}, \vec{k}) \to (\{l_i+1\}_{i \in \mathcal{D}}, \vec{k}) \to \ldots \to (\{l_i^* - 1\}_{i \in \mathcal{D}}, \vec{k}) \to (\vec{l}^*, \vec{k})$ or $(\vec{l}, \vec{k}) \to (\{l_i-1\}_{i \in \mathcal{D}}, \vec{k}) \to \ldots \to (\{l_i^* + 1\}_{i \in \mathcal{D}}, \vec{k}) \to (\vec{l}^*, \vec{k})$, where the probability of each transition is non-zero.

2) $(\vec{l}^*, \vec{k}) \longrightarrow (\vec{l}^*, \vec{h})$.

- First, we will prove that $p^{\vec{h}}_{\vec{l}*,\vec{k}} > 0$. Since $A_{i,t} = h_i - \max[0, k_i - R_{l_i^*}\Delta T] \geq 0$ when $R_{l_i^*}\Delta T \geq k_i$, we have $p^{h_i}_{l_i^*,k_i} > 0$ from (4.4) and $p^{\vec{h}}_{\vec{l}*,\vec{k}} > 0$ from (4.5).

- Then, we will prove that $p^{\vec{l}*}_{\vec{l}*,\vec{k},\vec{h}} > 0$. From (4.23), we only need to prove that $p^{l_i^*}_{(l_i^*,\vec{k},\vec{h})} > 0$. According to Theorem 4.1, this is equivalent to proving that $p^{l_i^*}_{(l_i^*,\theta_v,\theta_w)} > 0$, where $\vec{k} \in \mathcal{S}_{Q_i} \times \mathcal{S}^{\vec{i}}_{\theta_v}$ and $\vec{h} \in \mathcal{S}_{Q_i} \times \mathcal{S}^{\vec{i}}_{\theta_w}$. Since $\pi_{l_i^*|\theta_v} > 0$ and from (4.6), we need to prove that $p^{l_i^*,l_i^*}_{(\theta_v,\theta_w)} > 0$. According to (4.17), since we have proved that $p^{(l_i^*,l_i^*+1)}_{(\theta_v,\theta_v)} > 0$, $p^{(l_i^*,l_i^*-1)}_{(\theta_v,\theta_v)} > 0$, and $p^{(l_i^*,l_i^*)}_{(\theta_v,\theta_v)} > 0$ in the above discussion, we only need to show that at least one of three probabilities $\hat{p}^{l_i}_{(l_i^*+1,\theta_v,\theta_w)}$, $\hat{p}^{l_i^*}_{(l_i^*-1,\theta_v,\theta_w)}$, and $\hat{p}^{l_i^*}_{(l_i^*,\theta_v,\theta_w)}$ is non-zero. From Lemma 4.1, we have $\hat{p}^{l_i^*}_{(l_i^*,\theta_v,\theta_w)}$ is always non-zero irrespective of the relationship between θ_v and θ_w. Therefore, we can prove that $p^{\vec{l}*}_{\vec{l}*,\vec{k},\vec{h}} > 0$.

- Therefore, the transition from $(\vec{l}*,\vec{k})$ to $(\vec{l}*,\vec{h})$ has non-zero probability from (4.3).

3) $(\vec{l}*,\vec{h}) \longrightarrow (\vec{n},\vec{h})$. The proof is the same with (1).

Combining (1), (2) and (3), we can prove that there exits a multi-transition path with non-zero probability from state $(\vec{l},\vec{k})$ to $(\vec{n},\vec{h})$, i.e, $(\vec{l},\vec{k}) \longrightarrow (\vec{l}*,\vec{k}) \rightarrow (\vec{l}*,\vec{h}) \longrightarrow (\vec{n},\vec{h})$, where $R_{l_i^*}\Delta T \geq k_i$. Since $K \leq R_L\Delta T$, there always exists such l_i^* that satisfies this condition.

Lemma 4.3. *The Markov chain $(\vec{H}_t, \vec{Q}_t)$ is homogeneous and positive recurrent, if $K \leq R_L\Delta T$.*

Proof. Since the transition probability matrix $\mathbf{P}$ is independent of the time slot t, the Markov chain is homogeneous [27]. From Theorem 3.3 in [27], $(\vec{H}_t, \vec{Q}_t)$ is positive recurrent, since it has finite state space $(L \times (K+1))^D$ and is irreducible from Lemma 4.2.

From Theorem 3.1 in [27], Theorem 4.3 is valid if and only if the Markov chain $(\vec{H}_t, \vec{Q}_t)$ is irreducible, homogeneous and positive recurrent, which have been proved in Lemma 4.2 and Lemma 4.3.

References

1. L. Lei, Y. Zhang, X. Shen, C. Lin, and Z. Zhong (2013) Performance Analysis of Device-to-Device Communications with Dynamic Interference using Stochastic Petri Nets. IEEE Trans. Wireless Commun, 12(12):6121–6141
2. L. Lei, Z. Zhong, C. Lin et al (2012) Operator controlled device-to-device communications in LTE-advanced networks. IEEE Wireless Mag 19(3):96–104

3. G. Fodor, E. Dahlman, G. Mildh et al (2012) Design aspects of network assisted device-to-device communications. IEEE Commun. Mag 50(3):170–177
4. M. Scott Corson, R. Laroia, J. Li et al (2010) Toward proximity-aware internetworking. IEEE Wireless Mag 17(6):26–33
5. K. Doppler, M. Rinne, C. Wijting, C. B. Ribeiro et al (2009) Device-to-device communication as an underlay to LTE-Advanced Networks. IEEE Commun 47(12):42–49
6. C.H. Yu, K. Doppler, C. B. Ribeiro et al (2011) Resource Sharing Optimization for Device-to-Device Communication Underlaying Cellular Networks. IEEE Trans. Wireless Commun 10(8):2752–2763
7. X. Wu et al (2010) FlashLinQ: A Synchronous Distributed Scheduler for Peer-to-Peer Ad Hoc Networks. IEEE/ACM Trans. Networking 21(4):1215–1228
8. H. Wang, X. Chu (2012) Distance-constrained resource-sharing criteria for device-to-device communications underlaying cellular networks. Electronics Letters 48(9):528–530
9. M. Jung, K. Hwang, S. Choi (2012) Joint Mode Selection and Power Allocation Scheme for Power-Efficient Device-to-Device (D2D) Communication. Paper presented at the IEEE 75th Vehicular Technology Conference Spring, 6–9 May 2012
10. M. Zulhasnine, C. Huang, A. Srinivasan (2010) Efficient Resource Allocation for Device-to-Device Communication Underlaying LTE Network. Paper presented at the IEEE 6th Wireless Communications and Networking Conference, 11–13 Oct 2010
11. H. Xing, and S. Hakola (2010) The investigation of power control schemes for a device-to-device communication integrated into OFDMA cellular system. Paper presented at the IEEE 21st International Symposium on Personal Indoor and Mobile Radio Communications, 26–30 Sept 2010
12. C.-H. Yu, O. Tirkkonen, K. Doppler et al (2009) Power Optimization of Device-to-Device Communication Underlaying Cellular Communication. Paper presented at the IEEE International Conference on Communications, 14–18 June 2009
13. G. Fayolle, R. Iasnogorodski (1979) Two coupled processors: the reduction to a Riemann-Hilbert problem. Z. Wahr. verw. Geb 47:324–351
14. A. G. Konheim, I. Meilijson, A. Melkman (1981) Processor-sharing of two parallel lines. J. Appl. Probab 18(4):952–956
15. F. Guillemin, D. Pinchon (2004) Analysis of generalized processor-sharing systems with two classes of customers and exponential services. J. Appl. Probab 41(3):832–858
16. T. Bonald, S. Borst, N. Hegde, A. Proutiere (2004) Wireless data performance in multi-cell scenarios. Paper presented at the ACM Sigmetrics, 2004
17. T. Bonald, S. Borst, A. Proutiere (2005) Inter-cell scheduling in wireless data networks. Paper presented at the European Wireless Conference, 2005
18. B. Rengarajan, C. Caramanis, d. G. Veciana (2008) Analyzing queuing systems with coupled processors through semdefinite programming. http://users.ece.utexas.edu/~gustabo/papers/SdpCoupledQs.pdf
19. Rengarajan B, Veciana d. G (2011) Practical Adaptive User Association Policies for Wireless Systems with Dynamic Interference. IEEE/ACM Trans. Networking 19(6):1690–1703
20. H. S. Wang, N. Moayeri (1995) Finite-state Markov channel-A useful model for radio communication channels. IEEE Trans. Veh. Technol. 44:163–171
21. Q. Liu, S. Zhou, G. B. Giannakis (2005) Queueing with adaptive modulation and coding over wireless links: cross-layer analysis and design. IEEE Trans. Wireless Commun 50(3):1142–1153
22. L. Lei, C. Lin, J. Cai, X. Shen (2009) Performance analysis of opportunistic wireless schedulers using Stochastic Petri Nets. IEEE Trans. Wireless Commun 7(4):5461–5472
23. L. Le, E. Hossain (2008) Tandem Queue Models with Applications to QoS Routing in Multihop Wireless Networks. IEEE Trans. Mobile Computing 7(8):1025–1040
24. 3GPP TS 36.213, V10.3.0 (2011) Technical specification group radio access network; Evolved Universal Terrestrial Radio Access; Physical Layer Procedures (Release 10)

25. Selection procedures for the choice of radio transmission technologies of the UMTS. 3GPP TR 30.03U, version 3.2.0, 1998
26. ITU-R M.1225 (1997) Guidelines for the Evaluation of Radio Transmission Technologies (RTTs) for IMT-2000
27. P. Bremaud, Markov Chains: Gibbs Fields, Monte Carlo Simulation, and Queues. Springer-Verlag, 1999

Chapter 5
Packet Level Wireless Channel Model for OFDM System Using SHLPNs

In Chap. 2, we have introduced the SHLPNs and the compound marking technique for model aggregation. In this chapter, we adopt this technique to form a wireless channel model for OFDM systems in order to simplify the cross-layer performance analysis of modern wireless systems [1]. Compared with existing FSMC model whose state space grows exponentially with the number of OFDM subchannels, our proposed SHLPN model uses state aggregation technique to deal with this problem. Closed-form expressions to calculate the transition probabilities among the compound markings of the SHLPN model are provided. When applied to derive the performance measures for OFDM system in terms of the average throughput, average delay, and packet dropping probability, the SHLPN model can accurately capture the correlated time-varying nature of wireless channels. Simulation is performed to show that the numerical results offered by the proposed model are more accurate compared with other simplified channel models for avoiding state space complexity.

5.1 Packet Level Wireless Channel Model

Compared with wireline networks, the nature of wireless systems is dominated by the channel between antennas. It is critical for networking researchers and engineers to capture the channel characteristics in their cross-layer performance model, which are subject to complex phenomena such as multipath fading, Doppler, and time-dispersive effects introduced by the wireless propagation. Although wireless channel modeling for physical layer has been a very active area [2, 3], it is too complex to be incorporated into the cross-layer models for performance analysis and optimization.

As a link between the physical layer and higher layers, the wireless channel can be modeled as a first-order FSMC, which is suitable to be integrated into the cross-layer performance models. The FSMC modeling of flat-fading channels has been

© The Author(s) 2015   79
L. Lei et al., *Stochastic Petri Nets for Wireless Networks*, SpringerBriefs
in Electrical and Computer Engineering, DOI 10.1007/978-3-319-16883-8_5

studied for more than half a decade, ranging from the simple two-state GE channel to the more complex finite-state Markov channel [4]. Specifically, the SNR region is divided into multiple non-overlapping consecutive regions for a FSMC, and the channel of any wireless link is in a certain state if its received SNR falls in the corresponding region. The steady-state probability of each state can be derived by integrating the pdf of the SNR over the corresponding region. When the rate of temporal channel variations is relatively slow and the number of states is not so high, the state transition probability can be approximated by Level-Crossing Rate (LCR), which assumes that the channel state either stays in the same state or it transits to its immediate neighboring states from two consecutive time slots. The FSMC model is widely adopted in the cross-layer performance analysis of narrow-band wireless systems using AMC scheme [5–10], where the service process can be modeled as a MMDP. For example, an analytical framework is presented in [5], which constructs a two-dimensional Discrete Time Markov Chain containing both the queue and channel states to analyze the joint effects of finite-length queuing and AMC schemes on several QoS metrics.

The wireless channels of the next-generation wireless networks are wideband and frequency-selective. For example, the maximum bandwidth supported by 3GPP LTE and LTE-Advanced systems are 20 and 100 MHz, respectively. The frequency-selective fading channels can be turned into multiple parallel flat fading channels by OFDM, which is a physical-layer multi-carrier technology for effectively combating Inter-Symbol Interference (ISI). Therefore, the FSMC model for OFDM system can be represented as multiple parallel first-order sub-FSMCs, which may be correlated or independent from each other depending on the correlated bandwidth. In [11], an FSMC model for OFDM systems over Nakagami-m fading channel is introduced, where the correlation between different subbands are modeled by the LCR of frequency response in frequency domain. Although it is theoretically feasible to expand the analytical framework in [5] using such an FSMC model for cross-layer performance analysis of OFDM system, the exponential growth of the state space with the number of subchannels forbids its practical application.

Due to the complexity of the state space, existing research in cross-layer performance analysis and optimization generally makes some types of simplified assumptions when dealing with OFDM system. In the first type of simplification, the two-state ON-OFF channel model is used instead of the FSMC model with multiple states, which cannot accurately reflect the AMC scheme in practical wireless systems [12–14]. In the second type of simplification, the first-order and sometimes second-order statistics of the service process are used so that there is no need to calculate the transition probabilities of the FSMC. For example, the OFDM-Time Division Multiple Access (TDMA) and OFDMA systems are formulated as M/G/1 queuing models for packet-level performance analysis in [15], where only up to the second moments of the service process are needed to derive the average packet delay. However, more accurate analysis can be achieved if the service process is modeled as MMDP instead of general distribution. In the flow-level model, it is shown that an approximate analysis can be performed in a time-scale decomposable regime, where the time scale of the data file transmission time is much longer than that of the service process fluctuation [16]. In this case, the random fluctuations in the service rate become negligible, and a simple constant-rate service process can be applied. However, this simplification causes much inaccuracy for packet-level model, where channel fluctuations give rise to random time-varying service rates. In the third type of simplification, it is assumed that the channel state changes independently in consecutive time slots, where the state transition probability reduces to the steady-state probability [17, 18]. However, as the fading channel amplitude varies with time in a correlated manner, this simplification will cause some inaccuracy.

In this chapter, we first present a SPN model for OFDM wireless channel which is isomorphic with the FSMC model and can accurately capture the correlated time-varying nature of wireless channels. Moreover, we use the SHLPN formalism introduced in Sect. 2.3 for state aggregation to deal with the state space explosion problem. Specifically, we assume that the multiple parallel flat fading channels of the OFDM systems are stochastically identical and independent from each other, which is a reasonable assumption when the channel width is larger than the coherence bandwidth of the environment and has been widely adopted in related research work [12, 17, 18]. Therefore, the corresponding SPN model is a homogeneous system with subsets of equivalent states, which can be grouped together in such a way that the SHLPN model of the system with compound markings contains only one compound state for each group of individual states in the original SPN model. In this case an equivalence relation exists among the SHLPN model with compound markings and the original SPN model, while the SHLPN model has a lower number of states. We also derive the closed-form expressions for state transition probabilities and steady-state probabilities of the compound states of the SHLPN model. In [19], it is proposed to use lumpable FSMC to reduce the expanded Markov channel of OFDM system to multiple smaller Markov channels while maintaining similar behavior. Although the concept of lumpable FSMC in [19] is similar to state aggregation by SHLPN in this chapter, it suffers from the following shortcomings. First, it is assumed that the channel state of at most one subchannel can be changed in a time slot, which is not true since the channel states of different subchannels change independently over time. Second, it does not provide closed-form expressions for state transition probabilities of the lumpable FSMC. Therefore, the state transition probabilities of the original expanded Markov channel have to be derived first, which is still limited by the exponentially increased state space.

5.2 SHLPN Model of OFDM Wireless Channel

5.2.1 SPN Model

We first consider a Rayleigh flat fading channel, which can be approximated by a FSMC model. Specifically, the SNR values are divided into L non-overlapping consecutive regions. For any $l \in \{1, \ldots, L\}$, the channel of a wireless link is in state l if its instantaneous SNR value γ falls within the l-th region $[\Gamma_{l-1}, \Gamma_l)$. Obviously, $\Gamma_0 = 0$ and $\Gamma_L = \infty$. We assume that time is slotted and each time slot has an equal length ΔT. Moreover, we assume that the channel state remains constant within a time slot and the channel state at time slot t only depends on the channel state at time slot $t - 1$. Therefore, the wireless channel can be represented by a FSMC $\{H_t\}$, $t = 0, 1, 2, \ldots$, where $H_t \in \{1, \ldots, L\}$.

The steady-state probability π_l of each state l can be derived by integrating the pdf of the SNR over the corresponding region $[\Gamma_{l-1}, \Gamma_l)$. The state transition probability $p_{l,n} := \Pr.\{H_{(t+1)} = n | H_t = l\}$ can be approximated by LCR, which assumes that from time slot t to $t + 1$, the FSMC either stays in the same state or it transits to its immediate neighboring states. For Rayleigh fading channel, π_l and $p_{l,n}$ can be derived from (3.35)–(3.40).

Now consider the more complex case of Rayleigh frequency-selective fading channel in OFDM system with N_F subchannels. In this case, since every subchannel of a wireless link is Rayleigh flat fading channel, the FSMC state for an OFDM wireless channel can be represented by a tuple $\mathbf{H}_t = \{H_t^{(m)} | m \in \{1, \ldots, N_F\}\}$, where $H_t^{(m)} \in \{1, \ldots, L\}$ denotes the local channel state of the wireless link on subchannel m. Specifically for any $l \in \{1, \ldots, L\}$, $H_t^{(m)} = l$ if the instantaneous SNR value on subchannel m falls within the l-th region $[\Gamma_{l-1}, \Gamma_l)$. Assume that $H_t^{(1)}, H_t^{(2)}, \ldots, H_t^{(N_F)}$ are i.i.d. random variables. Therefore, the steady-state probability $\pi_{\vec{l}}$ of each state $\vec{l} := \{l^{(1)}, \ldots, l^{(N_F)}\}$ can be derived as $\pi_{\vec{l}} = \prod_{m=1}^{N_F} \pi_{l^{(m)}}$, where $l^{(m)} \in \{1, \ldots, L\}$ denotes the local channel state of subchannel m and $\pi_{l^{(m)}}$ can be derived by (3.39). Moreover, the state transition probability $p_{\vec{l},\vec{n}} := \Pr.\{\mathbf{H}_{(t+1)} = \vec{n} | \mathbf{H}_t = \vec{l}\}$ can be derived as $p_{\vec{l},\vec{n}} = \prod_{m=1}^{N_F} p_{l^{(m)},n^{(m)}}$, where $p_{l^{(m)},n^{(m)}}$ denotes the local channel state transition probability of subchannel m and can be derived according to (3.35)–(3.40). Note that the above model also applies to the more general Nakagami-m frequency-selective fading channel, where every subchannel is a Nakagami-q (Hoyt) flat fading channel [20], whose analytical pdf and LCR are available to derive the corresponding steady-state probabilities and state transition probabilities as in (3.35)–(3.40). Moreover, the FSMC can also be extended to channel models which include the shadowing effect as well as the small-scale fading effect. For example, we can use the Nakagami-lognormal model in [21], where the analytical pdf and LCR in [21] for Nakagami-lognormal channels will then be used to establish the channel steady-state probability matrix and state transition probability matrix. Since the focus of this chapter is on how to reduce the state space of the FSMC model for OFDM system, which is not impacted by the considered physical layer channel model, we adopt the Rayleigh fading channel due to its simplicity.

DTSPN model of OFDM wireless channel is shown in Fig. 5.1, which is isomorphic with the FSMC model and composed of N_F identical subnets. The m-th subnet for any $m \in \{1, \ldots, N_F\}$ includes places $\{ph_l^{(m)}\}_{l=1}^{L}$ and transitions $\{t_{(l+1)l}^{(m)}\}_{l=1}^{L-1}$ and $\{t_{l(l+1)}^{(m)}\}_{l=1}^{L-1}$. $ph_l^{(m)}$ is a place for the local channel state l of

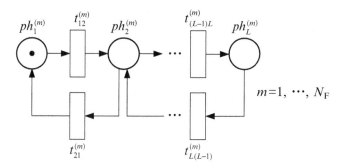

Fig. 5.1 The discrete time SPN model for OFDM wireless channel

subchannel m of the FSMC model, i.e., $M(ph_l^{(m)}) = 1$ if $H_t^{(m)} = l$. $\{t_{(l+1)l}^{(m)}\}_{l=1}^{L-1}$ and $\{t_{l(l+1)}^{(m)}\}_{l=1}^{L-1}$ are geometrically-distributed timed transitions for local channel state transitions of subchannel m. The firing probability of $t_{(l+1)l}^{(m)}$ (resp. $t_{l(l+1)}^{(m)}$) equals $p_{(l+1),l}^{(m)}$ (resp. $p_{l,(l+1)}^{(m)}$). When $t_{(l+1)l}^{(m)}$ (resp. $t_{l(l+1)}^{(m)}$) fires, the local channel state of subchannel m transits from $(l + 1)$ (resp. l) to l (resp. $(l + 1)$). Note that different from the SPN model which is isomorphic with a CTMC, the discrete time SPN model allows multiple firings at any time step. Specifically, N firings in a time step in our DTSPN model means that there are N subchannels with channel state transitions in this time step.

5.2.2 SHLPN Model

Although the above FSMC model or DTSPN model can be used to characterize the OFDM wireless channel, the cardinality of their state spaces are both L^{N_F}, which grows exponentially with the number of subchannels. The complexity of state space makes the above models hard to solve in practice. For example, there are 100 subchannels (when a subchannel corresponds to a resource block composed of 12 successive subcarriers) in an LTE system with 20 MHz bandwidth. Assume that there are 5 local channel states in total for each subchannel (i.e., $L = 5$), the set of markings in the DTSPN model consists of 5^{100} states, which is intractable by any MC solver [22].

In this section, we formulate the discrete time SHLPN model for the OFDM wireless channel as shown in Fig. 5.2, which is a scaled down version of the SPN model and has a smaller number of places, transitions and states as introduced in Sect. 2.3. In the SHLPN model each place and each transition stands for a set of places or transitions in the SPN model. The number of places is reduced to from $L \times N_F$ to L irrespective of the number of subchannels. For any $l \in \{1, \ldots, L\}$, the place ph_l stands for the set $\{ph_l^{(m)} | m = 1, \ldots, N_F\}$. Therefore, the number

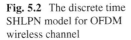

Fig. 5.2 The discrete time SHLPN model for OFDM wireless channel

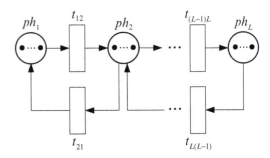

of tokens in place ph_l represents the number of subchannels with local channel state l, i.e., $M(ph_l) = \sum_{m=1}^{N_F} M(ph_l^{(m)}) = \sum_{m=1}^{N_F} \mathbf{1}(H_t^{(m)} = l)$. The number of transitions is reduced from $2(L-1) \times N_F$ to $2(L-1)$ irrespective of the number of subchannels. For any $l \in \{1, \dots, L-1\}$, the transition $t_{l(l+1)}$ (resp. $t_{(l+1)l}$) stands for the set $\{t_{l(l+1)}^{(m)} | m = 1, \dots, N_F\}$ (resp. $\{t_{(l+1)l}^{(m)} | m = 1, \dots, N_F\}$). The transition probability associated with every transition is related to the markings which enable that particular transition.

It is straightforward to see that there are a total of N_F tokens in the above SHLPN model, where each of the L places can contain $0, 1, \dots, N_F$ tokens. A compound marking (see Definition 2.1 of Sect. 2.3) can be represented as $\{k_1, k_2, \dots, k_L\}$ with $k_l = M(ph_l) \in \{0, 1, \dots, N_F\}$, i.e., k_l represents the number of tokens in place ph_l, and $\sum_{l=1}^{L} k_l = N_F$. Therefore, the total number of compound markings (states) in the state space is $\frac{(L+N_F-1)!}{N_F!(L-1)!}$, which grows much slower with the number of subchannels N_F compared with the exponentially increased state space size of the SPN model. The numbers of compound markings and individual markings with increasing values of N_F under different values of L are illustrated in Fig. 5.3. We define the mapping function from a compound marking to its index in the state space by $f_{id}(\cdot)$, i.e., $\hat{l} = f_{id}(\{k_1, k_2, \dots, k_L\})$, $\hat{l} \in \{1, \dots, \frac{(L+N_F-1)!}{N_F!(L-1)!}\}$. Let $\{\hat{H}_t\}_{t=0,1,2,\dots}$ be the underlying FSMC of the SHLPN, whose state equals the marking index, i.e., $\hat{H}_t = \hat{l} \in \{1, \dots, \frac{(L+N_F-1)!}{N_F!(L-1)!}\}$.

Since each compound marking corresponds to a subset of individual markings in the original SPN model, let $\mathcal{L} = f_{set}(\hat{l})$ represent the subset of individual markings corresponding to the compound marking with index $\hat{l}$, where $f_{set}(\cdot)$ is the mapping function from the compound marking index to the subset of individual markings. As an example, the compound marking (state) table of the SHLPN model and the corresponding individual markings in the SPN model is shown in Table 5.1, where there are 2 subchannels ($N_F = 2$) and 3 local channel states per subchannel ($L = 3$). There are 9 individual markings in the SPN model, which can be aggregated into 6 compound markings in the SHLPN model. Note that an individual marking represented as $\{M(ph_l^{(m)}) | l = 1, \dots, L; m = 1, \dots, N_F\}$ is equivalent with a state $\vec{l} = \{l^{(m)} | m = 1, \dots, N_F\}$ of the FSMC $\{\mathbf{H}_t\}_{t=0,1,2,\dots}$. For example, the

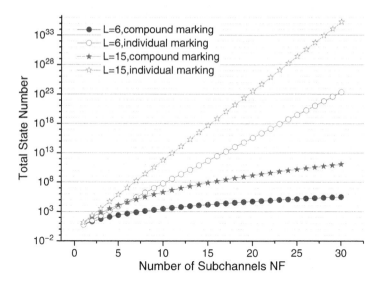

Fig. 5.3 The numbers of compound markings and individual markings with increasing number of subchannels N_{F} when $L = 6$ and $L = 15$, respectively

individual marking $\{1, 0, 0, 1, 0, 0\}$ in Table 5.1 is equivalent with state $\{1, 1\}$ of FSMC $\{\mathbf{H}_t\}_{t=0,1,2,\ldots}$.

Since the compound marking $\{k_1, k_2, \ldots, k_L\}$ means that there are currently k_l subchannels in local channel state $l \in \{1, \ldots, L\}$, and also because there are a total of N_{F} subchannels and the local channel states of all subchannels are i.i.d., the steady-state probability $\pi_{\hat{l}} := \lim_{t \to \infty} \mathrm{Pr}.(\hat{H}_t = \hat{l})$ of any compound marking $\{k_1, k_2, \ldots, k_L\}$ with index $\hat{l}$ can be derived as

$$\pi_{\hat{l}} = \prod_{l=1}^{L} C^{k_l}_{(N_{\mathrm{F}} - \sum_{l'=1}^{l-1} k_{l'})} (\pi_l)^{k_l}. \tag{5.1}$$

Based on (5.1), the steady-state probabilities of the compound markings of Table 5.1 is given in Table 5.2.

Theorem 5.1. *Denote by $p_{\hat{l}, \hat{n}}$ the probability of a transition from the compound marking with index $\hat{H}_t = \hat{l}$ to the compound marking with index $\hat{H}_{(t+1)} = \hat{n}$. The subset of individual markings corresponding to compound markings with indexes $\hat{l}$ and $\hat{n}$ are $\mathcal{L} = f_{\mathrm{set}}(\hat{l})$ and $\mathcal{N} = f_{\mathrm{set}}(\hat{n})$, respectively. Let $p_{\vec{l}, \vec{n}}$ be the probability of a transition from the individual marking $\mathbf{H}_t = \vec{l}$ to the individual marking $\mathbf{H}_{(t+1)} = \vec{n}$, where $\vec{l} \in \mathcal{L}$ and $\vec{n} \in \mathcal{N}$. By applying (2.3) in Sect. 2.3, we can*

Table 5.1 The compound markings

Compound markings				Individual markings						
Index ($\hat{l}$)	Tokens in places			Index	Tokens in places					
	$M(ph_1)$	$M(ph_2)$	$M(ph_3)$		$M(ph_1^{(1)})$	$M(ph_1^{(2)})$	$M(ph_1^{(3)})$	$M(ph_2^{(1)})$	$M(ph_2^{(2)})$	$M(ph_2^{(3)})$
1	2	0	0	1	1	0	0	1	0	0
2	0	2	0	2	0	1	0	0	1	0
3	0	0	2	3	0	0	1	0	0	1
4	1	1	0	4	1	0	0	0	1	0
5	1	0	1	5	0	1	0	1	0	0
6	0	1	1	6	1	0	0	0	0	1
				7	0	0	1	1	0	0
				8	0	1	0	0	0	1
				9	0	0	1	0	1	0

Table 5.2 The steady-state probabilities of compound markings

Compound markings	1({2, 0, 0})	2({0, 2, 0})	3({0, 0, 2})	4({1, 1, 0})	5({1, 0, 1})	6({0, 1, 1})
Steady-state probabilities	$(\pi_1)^2$	$(\pi_2)^2$	$(\pi_3)^2$	$2\pi_1\pi_2$	$2\pi_1\pi_3$	$2\pi_2\pi_3$

get the relation between the transition probability of compound markings and the transition probability of individual markings is:

$$p_{\hat{l},\hat{n}} = \sum_{\vec{n} \in \mathcal{N}} p_{\vec{l},\vec{n}}, \quad \forall \vec{l} \in \mathcal{L}. \tag{5.2}$$

Based on Theorem 5.1, the transition probabilities among the compound markings of Table 5.1 is given in Table 5.3. The value of $p_{l,n}$ can be derived according to (3.35)–(3.40). Since we assume that only state transitions among adjacent states are possible, the terms with underline equal 0 in Table 5.3.

Although Theorem 5.1 provides a way to calculate the transition probabilities among the compound markings, it is not convenient especially for large numbers of subchannels and local states. This is because all the individual states belonging to every compound state have to be enumerated and the transition probabilities between the individual states have to be calculated. In order to solve this problem, the following Lemma 5.1 provides the closed-form expression to directly calculate the transition probabilities among the compound markings of the SHLPN model for OFDM wireless channel.

Lemma 5.1. *Assume there are N_F subchannels and L local channel states per subchannel in the SHLPN model. The transition probability from compound marking $\{k_1, k_2, \ldots, k_L\}$ to compound marking $\{k_1', k_2', \ldots, k_L'\}$, where $\hat{l} = f_{id}(\{k_1, k_2, \ldots, k_L\})$ and $\hat{n} = f_{id}(\{k_1', k_2', \ldots, k_L'\})$, can be calculated as*

$$p_{\hat{l},\hat{n}} = \sum_{a_{1,2}=lb_{1,2}}^{ub_{1,2}} \cdots \sum_{a_{(L-1),L}=lb_{(L-1),L}}^{ub_{(L-1),L}} \prod_{l=1}^{L} C_{k_l}^{a_{l,(l-1)}} (p_{l,(l-1)})^{a_{l,(l-1)}}$$

$$C_{(k_l-a_{l,(l-1)})}^{a_{l,(l+1)}} (p_{l,(l+1)})^{a_{l,(l+1)}} (p_{l,l})^{(k_l-a_{l,(l-1)}-a_{l,(l+1)})}, \tag{5.3}$$

where $\forall l \in \{1, \ldots, L-1\}$

$$lb_{l,(l+1)} = \max[0, (\sum_{i=1}^{l}(k_i - k_i'))], \tag{5.4}$$

$$ub_{l,(l+1)} = \min[(k_l - a_{l,(l-1)}), (\sum_{i=1}^{l+1} k_i - \sum_{i=1}^{l} k_i'), k_{(l+1)}'], \tag{5.5}$$

Table 5.3 The transition probabilities among compound markings

Previous states	Post states					
	1({2,0,0})	2({0,2,0})	3({0,0,2})	4({1,1,0})	5({1,0,1})	6({0,1,1})
1({2,0,0})	$p_{1,1}p_{1,1}$	$p_{1,2}p_{1,2}$	$p_{1,3}p_{1,3}$	$2p_{1,1}p_{1,2}$	$2p_{1,1}p_{1,3}$	$2p_{1,2}p_{1,3}$
2({0,2,0})	$p_{2,1}p_{2,1}$	$p_{2,2}p_{2,2}$	$p_{2,3}p_{2,3}$	$2p_{2,1}p_{2,2}$	$2p_{2,1}p_{2,3}$	$2p_{2,2}p_{2,3}$
3({0,0,2})	$p_{3,1}p_{3,1}$	$p_{3,2}p_{3,2}$	$p_{3,3}p_{3,3}$	$2p_{3,1}p_{3,2}$	$2p_{3,1}p_{3,3}$	$2p_{3,2}p_{3,3}$
4({1,1,0})	$p_{1,1}p_{2,1}$	$p_{1,2}p_{2,2}$	$p_{1,3}p_{2,3}$	$p_{1,1}p_{2,2}+p_{1,2}p_{2,1}$	$p_{1,1}p_{2,3}+p_{1,3}p_{2,1}$	$p_{1,2}p_{2,3}+p_{1,3}p_{2,2}$
5({1,0,1})	$p_{1,1}p_{3,1}$	$p_{1,2}p_{3,2}$	$p_{1,3}p_{3,3}$	$p_{1,1}p_{3,2}+p_{3,1}p_{1,2}$	$p_{1,1}p_{3,3}+p_{1,3}p_{3,1}$	$p_{1,2}p_{3,3}+p_{1,3}p_{3,2}$
6({0,1,1})	$p_{2,1}p_{3,1}$	$p_{2,2}p_{3,2}$	$p_{2,3}p_{3,3}$	$p_{2,1}p_{3,2}+p_{2,2}p_{3,1}$	$p_{2,1}p_{3,3}+p_{2,3}p_{3,1}$	$p_{2,2}p_{3,3}+p_{2,3}p_{3,2}$

and

$$a_{(l+1),l} = \sum_{i=1}^{l} (k_l' - k_l) + a_{l,(l+1)}. \tag{5.6}$$

Moreover, $a_{1,0} = a_{L,(L+1)} = 0$ *and* $p_{1,0} = p_{L,(L+1)} = 0.$

The proof of Lemma 5.1 is given in the Appendix.

5.3   Example Application to Cross-Layer Performance Analysis of Cellular Downlink

5.3.1   Model Description

In this section, we show how the proposed SHLPN model for OFDM wireless channel can be applied to the cross-layer performance analysis of cellular downlink. Consider the downlink of a single-cell OFDM system, where a BS transmits data[1] to multiple mobile users. Note that if applied to multi-cell scenario and when inter-cell interference is not negligible due to frequency reuse between neighboring cells, the analytical pdf and LCR in Sect. 3.3.1 to derive the channel state transition probability matrix of every subchannel should be in terms of SINR instead of SNR. As long as the interference on all the subchannels or a group of subchannels are stochastically identical, our SHLPN model for state aggregation is still applicable. The derivation of the analytical pdf and LCR in terms of SINR shall be an interesting and challenging problem, which depends on the topology (e.g., femto, pico, relay) and inter-cell interference coordination mechanism. However, this problem is out of scope of this chapter.

The wireless channel is wideband and frequency-selective. We consider that each mobile user is assigned a fixed number of subchannels which remain unchanged during its service. Therefore, the service process of each mobile user depends only on its current channel and queue state, which makes it possible to separate the queuing dynamics of each user [18, 20]. This static scheduling policy is feasible for scenarios of best-effort users with no minimum QoS requirements. For other more general scenarios, it is interesting to study the performance of more sophisticated channel- and queue-aware multi-channel scheduling algorithms with the SHLPN model, e.g., using the technique of calculating the distribution of the maximum of a multi-dimensional Markov chain [23]. In the following discussion, we only focus on the queuing dynamics of one mobile user served by N_F subchannels, which can be applicable to others. The BS maintains a data buffer for the mobile

---

[1]The data can take units of bits or packets. The latter is appropriate when all the packets have fixed length.

user with a finite capacity of $N_Q < \infty$. We assume that the data arrival process is i.i.d. over time slots following general distribution with average arrival rate $E[A_t] = \lambda$, where $\mathrm{Pr}.(A_c = arr) > 0$ if $arr \geq 0$[2]. The transmission in the time is slot-by-slot based and each slot has an equal length ΔT. It is assumed that all channel conditions on every subchannel are available at the BS so that the AMC scheme can be applied. Specifically, the SNR values are divided into L non-overlapping consecutive regions. For any $l \in \{1, \ldots, L\}$, if the instantaneous SNR value $\gamma_t^{(m)}$ on subchannel m during time slot t falls within the l-th region $[\Gamma_{l-1}, \Gamma_l)$, the corresponding instantaneous data rate[3] $r_t^{(m)}$ on subchannel m is a fixed value R_l according to the selected modulation and coding scheme in this state, i.e., $r_t^{(m)} = R_l$, if $\gamma_t^{(m)} \in [\Gamma_{l-1}, \Gamma_l)$. The total instantaneous data rate of the wireless channel is the sum of the instantaneous data rate on every subchannel, i.e., $r_t = \sum_{m=1}^{N_F} r_t^{(m)}$.

The described system model can be formulated by the following queuing system, which consists of one queue fed with an arrival process that is i.i.d. over time slots following general distribution and one server. There is a FSMC $\hat{H}_t$ with total $\frac{(L+N_F-1)!}{N_F!(L-1)!}$ channel states. For any $\hat{l} \in \{1, \ldots, \frac{(L+N_F-1)!}{N_F!(L-1)!}\}$, if $\hat{H}_t = \hat{l} = f_{\mathrm{id}}(\{k_1, k_2, \ldots, k_L\})$, the queue is served at a deterministic service rate $r_{\hat{l}} = \sum_{l=1}^{L} k_l R_l$, which is a non-negative integer. Therefore, the queue is served according to an $\frac{(L+N_F-1)!}{N_F!(L-1)!}$-state MMDP.

Let Q_t denote the length of the queue at the beginning of time slot t. If Q_t is less than r_t during time slot t, padding bits shall be transmitted along with the data. Arriving data are placed in the queue throughout the time slot t and can only be transmitted during the next time slot $t + 1$. If the queue length reached the buffer capacity N_Q, the subsequent arriving data will be dropped. According to the above assumption, the queuing process evolves as follows:

$$Q_{t+1} = \min \left[N_Q, \max[0, Q_t - r_t] + A_t \right]. \tag{5.7}$$

The system behavior of the above queuing model can be accurately represented by the two-dimensional Markov chain $\{\hat{H}_t, Q_t\}_{t=0,1,2,\ldots}$. Let $p_{(\hat{l},q),(\hat{n},h)}$ be the transition probability from state $(\hat{l}, q)$ to state $(\hat{n}, h)$ of the Markov chain. Then,

$$p_{(\hat{l},q),(\hat{n},h)} = \mathrm{Pr}.\{Q_{(t+1)} = h | \hat{H}_t = \hat{l}, Q_t = q\} p_{\hat{l},\hat{n}} = v_{q,h}^{\hat{l}} p_{\hat{l},\hat{n}}, \tag{5.8}$$

---

[2]The number of assigned subchannels, the data buffer capacity, and the average arrival rate can be different across different mobile users.

[3]The instantaneous data rate can take units of bits/slot or packets/slot. The latter is appropriate when all the packets have fixed length and the achievable data rates are constrained to integral multiples of the packet size.

where $p_{\hat{l},\hat{n}}$ denotes the transition probability of the FSMC from state $\hat{l}$ to $\hat{n}$ and can be calculated according to Lemma 5.1.

According to (5.7) we have,

$$
v_{q,h}^{\hat{l}} = \begin{cases}
\text{Pr.}(A_t = h - q + r_{\hat{l}}) & q \geq r_{\hat{l}}, h \neq N_Q, \\
\text{Pr.}(A_t = h) & q < r_{\hat{l}}, h \neq N_Q, \\
\text{Pr.}(A_t \geq N_Q - q + r_{\hat{l}}) & q \geq r_{\hat{l}}, h = N_Q, \\
\text{Pr.}(A_t \geq N_Q) & q < r_{\hat{l}}, h = N_Q,
\end{cases}
\tag{5.9}
$$

where $\text{Pr.}(A_t = arr)$ is known with mean λ.

Define the state transition probability matrix $\mathbf{P} = [p_{(\hat{l},q),(\hat{n},h)}]$. Define the steady-state probability $\pi_{\hat{l},q} \equiv \lim_{t\to\infty} \text{Pr.}\{\hat{H}_t = l, Q_t = q\}$ and the vector $\boldsymbol{\pi} = (\pi_{1,0}, \ldots, \pi_{1,N_Q}, \ldots, \pi_{\frac{(L+N_F-1)!}{N_F!(L-1)!},0}, \ldots, \pi_{\frac{(L+N_F-1)!}{N_F!(L-1)!},N_Q})$. Then, the stationary distribution of the ergodic process $\{\hat{H}_t, Q_t\}_{t=0,1,\ldots}$ can be uniquely determined from the balance equations

$$
\boldsymbol{\pi} = \boldsymbol{\pi}\mathbf{P}, \quad \boldsymbol{\pi}\mathbf{e} = 1,
\tag{5.10}
$$

where $\mathbf{e}$ is the unity vector of dimension $\frac{(L+N_F-1)!}{N_F!(L-1)!} \times (N_Q + 1)$ and $\boldsymbol{\pi}$ can be derived as the normalized left eigenvector of $\mathbf{P}$ corresponding to eigenvalue 1. Given $\boldsymbol{\pi}$, the performance metrics such as the average queue length, the mean throughput, the average delay and the dropping probability can be derived similar to (14)–(19) in [24].

5.3.2 Numerical Results

We consider a wireless network employing adaptive M-ary quadrature amplitude modulation (M-QAM) with convolutional coding which has six channel states for all transmission links. The SNR thresholds for the channel states are given in Table II of [5]. We assume the Rayleigh fading channel and the number of packets transmitted in a time slot under different channel states, i.e., R_l with $l = 1, 2, 3, 4, 5, 6$ are set to 0, 1, 2, 3, 6, 9, respectively. The carrier frequency and the time slot duration ΔT are set to 2 GHz and 1 ms, respectively. The velocity of the terminals is set to be 3 km/h so that the Doppler frequency becomes 5.56 Hz. The mean SNR is 0 dB. We let the buffer size $K = 10$ packets, where the packet length $B = 1080$ bits.

We numerically solve the analytical model described in Sect. 5.3.1 and compare the performance measures with those obtained by discrete-event simulations of a downlink OFDM system. Moreover, we also perform numerical experiments which replace the SHLPN channel model with two other simplified channel models widely used in existing literature on OFDM system performance analysis [16, 17]:

- Constant-rate model, where the wireless channel is assumed to transmit at constant rate which equals to the expected channel transmission capability under all channel states.
- Memoryless model, where the channel state is assumed to change independently in consecutive time slots and the state transition probabilities reduces to the steady-state probabilities.

All three numeric methods and simulation are implemented in Matlab. We increase the subchannel number N_F from 1 to 5. In the simulation, we generate N_F i.i.d Rayleigh fading channels by the Jakes Model using a U-shape Doppler power spectrum [25] for every subchannel. In each simulated time slot, packets arrive according to Poisson distribution with mean $\lambda = 4$ packets/slot. If the queue in the current time slot is non-empty, we derive the SNR values of every subchannel according to the channel gains generated by the Jakes Model. The corresponding transmission rates on every subchannel in this time slot can thus be derived according to Table II of [5]. The simulations are run over 10^5 time slots and the time-average performance measures are obtained. Figure 5.4 shows the

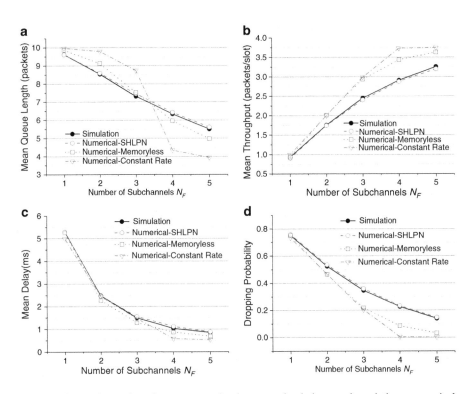

Fig. 5.4 Comparison of performance metrics between simulation result and three numerical results based on SHLPN, memoryless, and constant rate channel models, respectively. (**a**) Mean queue length, (**b**) mean throughput (packets/slot), (**c**) mean delay (ms), (**d**) dropping probability

average queue length, average throughput, average delay, and dropping probability obtained by the three numeric methods and simulation. It can be observed that the numeric results using SHLPN channel model is very close to the simulation results, while there are obvious differences between the numeric results using the two simplified channel models and the simulation results. The estimated performance by both Memoryless model and Constant-rate model are optimistic compared with the simulation results and the estimated performance by the SHLPN model in terms of average throughput, mean delay and dropping probability. The Memoryless model represents the scenario where the fading speed is extremely fast so that the time-correlation of the wireless channel between two consecutive time slots tends to zero. Therefore, we can reasonably conjecture that higher fading speed improves the average packet-level performance. Similar conclusions have been drawn for the worst-case packet-level performance [26] and average flow-level performance [27]. Moreover, since the Constant-rate model provides the best performance among the three numeric methods, we can see that the channel variation due to fast fading has negative effects on the packet-level performance. It can also be observed that the performance gaps between the three numeric methods increase with the number of subchannels from 1 to 4. When the number of subchannels increases to 4, the mean throughput of the Constant-rate server almost reaches 4 packets/slot, which equals the mean arrival rate. Its dropping probability becomes almost zero at the same time. Therefore, the mean throughput of the Constant-rate server stops increasing when the subchannel number further increases to 5. Note that it takes 5 subchannels for the Memoryless model to reach this saturated throughput and even more subchannels for the SHLPN model.

5.4 Summary

In this chapter, we have developed a wireless channel model using SHLPN formalism for the cross-layer performance analysis of OFDM system. The main reason for adopting SHLPN in channel modeling is to deal with the state space explosion problem of existing FSMC model by the state aggregation technique. Specifically, there is an equivalence relation between the SHLPN model and the FSMC model, while the former has a lower number of states. Closed-form expressions for state transition probabilities and steady-state probabilities of the compound states of the SHLPN model are derived. Finally, we apply the SHLPN model to cross-layer performance analysis of a downlink OFDM system and obtain performance measures such as average delay and dropping probability, etc.. The numerical results are validated by simulation, and it is shown that performance analysis based on the SHLPN model is more accurate compared to those based on the two simplified channel models used in existing literature. Note that the main results in the chapter are not restricted to OFDM systems, but can also be applied to other parallel broadcast systems including Multiple-Input Multiple-Output (MIMO) and carrier aggregation.

Appendix: Proof of Lemma 5.1

According to Theorem 5.1, the transition probability $p_{\hat{l},\hat{n}}$ from compound marking $\{k_1, k_2, \ldots, k_L\}$ with index $\hat{l}$ to compound marking $\{k'_1, k'_2, \ldots, k'_L\}$ with index $\hat{n}$ equals the sum of the transition probabilities from any one of the individual markings $\vec{l} \in \mathcal{L}$ to all of the individual markings $\vec{n} \in \mathcal{N}$. Given an individual marking $\vec{l} \in \mathcal{L}$, we not only know that there are k_l subchannels in local channel state l for any $l \in \{1, \ldots, L\}$, but also the specific subset of subchannels in each local state l. In order to calculate $p_{\hat{l},\hat{n}}$, we need to enumerate all the events that lead to the number of subchannels in local channel state l transited to k'_l for any $l \in \{1, \ldots, L\}$.

As illustrated in Fig. 5.5, let $a_{l,(l+1)}$ and $a_{(l+1),l}$ be the number of subchannels that transit from local channel state l to $(l+1)$ and vice versa, respectively. With an individual marking $\vec{l} \in \mathcal{L}$, and assuming that the values of $a_{l,(l+1)}$ and $a_{(l+1),l}$ are known for all $l \in \{1, \ldots, L-1\}$, the probability of this event can be derived as

$$\prod_{l=1}^{L} C_{k_l}^{a_{l,(l-1)}} (p_{l,(l-1)})^{a_{l,(l-1)}} C_{(k_l-a_{l,(l-1)})}^{a_{l,(l+1)}} (p_{l,(l+1)})^{a_{l,(l+1)}} (p_{l,l})^{(k_l-a_{l,(l-1)}-a_{l,(l+1)})},$$

$$(5.11)$$

where we set $a_{1,0} = a_{0,1} = a_{L,(L+1)} = a_{(L+1),L} = 0$.

Therefore, in order to derive $p_{\hat{l},\hat{n}}$, we only need to find all the possible values of $a_{l,(l+1)}$ and $a_{(l+1),l}$ that result in k_l transited to k'_l for every $l \in \{1, \ldots, L-1\}$, and calculate the sum probabilities of these events. For this purpose, the following equation needs to be true for any $l \in \{1, \ldots, L-1\}$:

$$k_l - a_{l,(l-1)} - a_{l,(l+1)} + a_{(l-1),l} + a_{(l+1),l} = k'_l. \qquad (5.12)$$

Adding the first l equations of (5.12) together, we can get (5.6) in Lemma 5.1, which establishes the relationship between $a_{l,(l+1)}$ and $a_{(l+1),l}$. Now, we only need to find the upper and lower bounds of $a_{l,(l+1)}$.

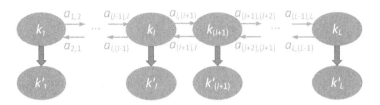

Fig. 5.5 Illustration of the state transition of compound markings from $\{k_1, k_2, \ldots, k_L\}$ to $\{k'_1, k'_2, \ldots, k'_L\}$. k_l is the number of subchannels originally in local state l. $a_{l,(l+1)}$ and $a_{(l+1),l}$ are the number of subchannels that transit from local channel state l to $(l+1)$ and vice versa, respectively. After the transition, the number of subchannels in local state l becomes k'_l

We first derive the upper bound of $a_{l,(l+1)}$. Since the number of subchannels transited from state l to other states must not be larger than the number of subchannels originally in state l, and the number of subchannels transited to state l from other states must not be larger than the number of subchannels in state l after the transition, the following four inequalities related to $a_{l,(l+1)}$ and $a_{(l+1),l}$ must hold:

$$\begin{cases} a_{l,(l+1)} + a_{l,(l-1)} \leq k_l, \\ a_{(l+1),l} + a_{(l+1),(l+2)} \leq k_{(l+1)}, \\ a_{(l+1),l} + a_{(l-1),l} \leq k'_l, \\ a_{l,(l+1)} + a_{(l+2),(l+1)} \leq k'_{(l+1)}. \end{cases} \tag{5.13}$$

Since we need to enumerate all the possible values of $a_{l,(l+1)}$ for every $l \in \{1, \ldots, L-1\}$, we start from $l = 1$ and proceed to increasing values of l in sequence. Therefore, when we set the upper bounds for $a_{l,(l+1)}$ and $a_{(l+1),l}$, we can assume that the values of $a_{l,(l-1)}$ and $a_{(l-1),l}$ are known, and ignore the values of $a_{(l+1),(l+2)}$ and $a_{(l+2),(l+1)}$ since the upper bounds of these two values will be set as a function of the values of $a_{l,(l+1)}$ and $a_{(l+1),l}$. Combining with (5.6), we find that the first and the third inequalities are equivalent. Therefore, the above four inequalities becomes:

$$\begin{cases} a_{l,(l+1)} \leq k_l - a_{l,(l-1)}, \\ a_{l,(l+1)} \leq \sum_{i=1}^{(l+1)} k_i - \sum_{i=1}^{(l)} k'_i, \\ a_{l,(l+1)} \leq k'_{(l+1)}, \end{cases} \tag{5.14}$$

which establishes the upper bound of $a_{l,(l+1)}$ by taking the minimum value of the right hand sides of the above three inequalities and results in (5.5) in Lemma 5.1.

Next, we derive the lower bound of $a_{l,(l+1)}$ from the fact that $a_{l,(l+1)} \geq 0$ and $a_{(l+1),l} \geq 0$. Therefore, we have

$$\begin{cases} a_{l,(l+1)} \geq 0, \\ a_{l,(l+1)} \geq \sum_{i=1}^{(l)}(k_i - k'_i), \end{cases} \tag{5.15}$$

which establishes the lower bound of $a_{l,(l+1)}$ by taking the maximum value of the right hand sides of the above two inequalities and results in (5.4) in Lemma 5.1.

Now armed with the upper and lower bounds of $a_{l,(l+1)}$, we can enumerate all the possible values of $a_{l,(l+1)}$ for every $l \in \{1, \ldots, L-1\}$ and calculate the sum probabilities of these events, which results in (5.3) in Lemma 5.1 and completes the proof.

References

1. L. Lei, H. Wang, C. Lin and Z. Zhong (2014) Wireless Channel Model using Stochastic High-Level Petri Nets for Cross-Layer Performance Analysis in Orthogonal Frequency-Division Multiplexing System, IET Communications 8(16):2871–2880
2. X. Cheng et al (2013) Wideband Channel Modeling and ICI Cancellation for Vehicle-to-Vehicle Communication Systems. IEEE Journal on Selected Areas in Communications 31(9):434–448
3. X. Cheng et al (2012) Cooperative MIMO Channel Modeling and Multi-link Spatial Correlation Properties. IEEE Journal on Selected Areas in Communications 30(2):388–396
4. P. Sadeghi, R. A. Kennedy, P. B. Rapajic et al (2008) Finite-state markov modeling of fading channels - a survey of principles and applications. IEEE Signal Processing Magazine 25(5): 57–80
5. Q. Liu, S. Zhou, G. B. Giannakis (2005) Queueing with adaptive modulation and coding over wireless links: cross-layer analysis and design. IEEE Trans. Wireless Commun 50(3): 1142–1153
6. K. Zheng, Y. Wang, L. Lei, W. Wang (2010) Cross-layer queuing analysis on multihop relaying networks with adaptive modulation and coding. IET Communications 4(3):295–302
7. S. Zhou. K. Zhang, Z. Niu, Yang Yang (2008) Queuing Analysis on MIMO Systems with Adaptive Modulation and Coding. Paper presented at the IEEE International Conference on Communications (ICC), 19–23 May 2008
8. K. Zheng, F. Liu, L. Lei, C. Lin et al (2013) Stochastic performance analysis of a wireless finite-state Markov channel. IEEE Transactions on Wireless Communications 12(2):782–793
9. J. Ramis, G. Femenias (2013) Cross-Layer QoS-Constrained Optimization of Adaptive Multi-Rate Wireless Systems using Infrastructure-Based Cooperative ARQ. IEEE Transactions on Wireless Communications 12(5):2424–2435
10. H. Chen, H. C. B. Chan et al (2013) QoS-Based Cross-Layer Scheduling for Wireless Multimedia Transmissions with Adaptive Modulation and Coding. IEEE Transactions on Communications 61(11):4526–4538
11. L. Cai, X. Shen, J. W. Mark (2009) Multimedia Services in Wireless Internet: Modeling and Analysis. John Wiley & Sons, 2009
12. B. Ji, G. R. Gupta, X. Lin, N. B. Shroff (2013) Performance of low-complexity greedy scheduling policies in multi-channel wireless networks: optimal throughput and near-optimal delay. Paper presented at the IEEE Internatioanl Conference on Computer Communications (INFOCOM), 14–19 April 2013
13. S. Bodas, S. Shakkottai, L. Ying, R. Srikant (2010) Low-complexity scheduling algorithms for multi-channel downlink wireless networks. Paper presented at the IEEE Internatioanl Conference on Computer Communications (INFOCOM), 14–19 March 2010
14. S. Kittipiyakul, T. Javidi (2009) Delay-optimal server allocation in multiqueue multiserver systems with time-varying connectivities. IEEE Trans. Information Theory 55(5):2319–2333
15. Y. J Chang, F. T Chien, C. Kuo (2007) Cross-layer QoS analysis of opportunistic OFDM-TDMA and OFDMA networks. IEEE Journal on Selected Areas in Commun 25(4):657–666
16. L. Lei, C. Lin, J. Cai et al (2008) Flow-level Performance of Opportunistic OFDM-TDMA and OFDMA Networks. IEEE Trans. Wireless Commun 7(12):5461–5472
17. Y. Cui, V. K. N. Lau (2010) Distributive Stochastic Learning for Delay-Optimal OFDMA Power and Subband Allocation. IEEE Trans. on Signal Process 58(9):4848–4858
18. L. Le, E. Hossain (2008) Tandem queue models with applications to QoS routing in multihop wireless networks. IEEE Trans. Mobile Computing 7(8):1025–1040
19. T. K. Chee, C. C. Lim, and J. Choi (2006) Channel Prediction Using Lumpable Finite-State Markov Channels in OFDMA Systems. Paper presented at the IEEE 63rd Vehicular Technology Conference, 7–10 May 2006
20. C. S. Bae (2010) Modeling and Performance analysis of OFDM based multi-hop cellular networks. Ph.D thesis, KAIST, 2010

21. T. T. Tjhung, C. C. Chai (1999) Fade statistics in Nakagami-lognormal channels. IEEE Trans. on Commun 47(12):1769–1772
22. R. Schoenen, M. R. Salem, A. B. Sediq et al (2011) Multihop Wireless Channel Models suitable for Stochastic Petri Nets and Markov State Analysis. Paper presented at the IEEE 73rd Vehicular Technology Conference (VTC Spring), 15–18 May 2011
23. S. Bodas, S. Shakkottai, L. Ying et al (2011) Scheduling for small delay in multi-rate multi-channel wireless networks. Paper presented at the IEEE Internatioanl Conference on Computer Communications (INFOCOM), 10–15 April 2011
24. L. Lei, C. Lin, J. Cai, S. Shen (2009) Performance analysis of opportunistic wireless schedulers using Stochastic Petri Nets. IEEE Trans. Wireless Commun 7(4):2076–2087
25. ITU-R M.1225 (1997) Guidelines for the Evaluation of Radio Transmission Technologies (RTTs) for IMT-2000
26. M. Fidler (2006) A network calculus approach to probabilistic quality of service analysis of fading channels. Paper presented at the Global Telecommunications Conference, Nov. 27-Dec. 1 2006
27. T. Bonald, S. C. Borst, A. Proutiere (2004) How mobility impacts the flow-level performance of wireless data system. Paper presented at the 23rd AnnualJoint Conference of the IEEE Computer and Communications Societies, Hong Kong, 7–11 March 2004

Chapter 6
Conclusions and Outlook

In the previous chapters, we have reviewed the background and state-of-art research on using SPNs for performance modeling of wireless networks. In this chapter, we summarize the key conclusions and suggest possible future research directions.

6.1 Conclusions

This book is devoted to introducing SPNs to more wireless networking researchers, and sharing our experience on how to adopt this powerful high-level modeling formalism to tackle the unsolved problems in performance evaluation of wireless networks. The key conclusions are summarized as follows.

- In Chap. 1, we first showed that performance evaluation of wireless networks needs to be based on stochastic models instead of the much simpler deterministic models due to the stochastic nature of wireless channel conditions, traffic arrivals and underlying geometry. Moreover, the performance models of wireless networks can be broadly classified into packet level model and flow level model. Then, we showed that SPNs are a powerful high-level modeling formalism that have been widely adopted by researchers in computer science. Although the SPN models tend to result in Markov processes which have a large number of states, a rich theory of model decomposition and aggregation has been developed to tackle this problem. Finally, we reviewed the state-of-art research on SPNs for wireless networks.
- In Chap. 2, we introduced the basic background knowledge of SPNs and discussed two important techniques to deal with the well-known state space explosion problem—model decomposition and iteration and compound marking in SHLPNs.
- In Chap. 3, we adopted the model decomposition and iteration technique to study the performance of wireless opportunistic schedulers in multiuser systems under

© The Author(s) 2015
L. Lei et al., *Stochastic Petri Nets for Wireless Networks*, SpringerBriefs
in Electrical and Computer Engineering, DOI 10.1007/978-3-319-16883-8_6

a dynamic data arrival setting. Analytical results demonstrated that the multiuser diversity effect as observed in the infinite backlog scenario is only valid in the heavy traffic regime. The performance of the CA opportunistic schedulers is worse than that of the round robin scheduler in the light traffic regime.

- In Chap. 4, we adopted the model decomposition and iteration technique to study the performance of D2D communications with full frequency reuse between D2D links. The queuing behavior for such system can be modeled by a coupled-processor server, where the service rate at each queue vary over time as governed by the backlogged state of the other queues. The complex interaction between the various queues renders an exact analysis intractable in general and steady-state queue length distributions are known only for exponentially distributed service in two-queue systems. We showed that this non-trivial problem can be solved by using the model decomposition and iteration technique in SPNs.

- In Chap. 5, we adopted compound marking in SHLPNs to form a wireless channel model for OFDM systems in order to simplify the cross-layer performance analysis of modern wireless systems. The state space of the SHLPNs-based OFDM channel model no longer grows exponentially with the number of subchannels as the existing FSMC model does. The proposed channel model can be used for cross-layer performance analysis of OFDM systems.

6.2 Outlook

From the previous chapters, we know SPNs are an attractive modeling formalism with user-friendly graphical orientation, powerful and flexible modeling tool and solid mathematical basis. The example applications in this book only give a glimpse of what SPNs may bring to the modeling and performance analysis of wireless networks. As the future wireless networks, e.g., the 5G cellular networks, are expected to be a mixture of network tiers of different sizes, transmit powers, backhaul connections, different radio access technologies that are accessed by an unprecedented number of smart and heterogeneous wireless devices, it is very exciting to explore the development and application of SPNs to deal with the unsolved problems arising from the increasing system complexity and performance requirement. Several challenges and related potential future research directions are identified as follows.

1. *State space explosion problem in large-scale networks*: The scale of the future wireless networks will increase sharply in every dimension. For example, the massive MIMO communications introduce a very large number of service antennas; the M2M and D2D communications support many more connected devices; the ultra-dense small cells mean massive growth in the number of BSs; and the millimeter wave communications will utilize a very wide bandwidth. In general, solving the Markovian model for such large-scale networks is an

non-trivial problem as the state space grows exponentially with the system dimension. Therefore, it is very interesting to utilize the existing rich set of techniques in SPNs to deal with this problem.

2. *Non-Markovian SPNs for diverse traffic patterns*: Future wireless networks need to support a diverse set of services, applications and users. By 2020 there will be more than 30 times as much mobile internet traffic as there was in 2010. The traffic arrival patterns of many new applications differ from the exponential assumption adopted in Markovian models. On the other hand, several classes of non-Markovian SPNs have been developed which incorporate some non-exponential characteristics in their definition [1]. Therefore, an interesting research direction is to adopt these non-Markovian SPNs for the performance analysis of new applications with non-exponential inter-arrival time.

3. *Performance optimization based on SPN models*: SPNs are a high-level modeling formalism for performance evaluation of discrete event systems. The purpose of performance evaluation is to examine whether a given system will satisfy the performance requirements or not when its design is finished. On the other hand, it is also very important for a system to be designed in a way to achieve the optimal performance. For this purpose, SPNs can be combined with optimization theories to make the modeling phase easier. For example, Markov Decision Petri Nets (MDPNs) was proposed by Beccuti et al. in 2007 [2], which integrate SPNs and Markov Decision Process (MDP) in order to model and analyze distributed systems with probabilistic and non deterministic features. It is of great interest to investigate how to apply the SPN-based optimization theories in wireless networks.

References

1. A. Bobbio, A. Puliafito, M. Telek, K.S. Trivedi (1998) Recent Developments in Non-Markovian Stochastic Petri Nets. Journal of Circuits Systems and Computers 8(1):119–158
2. M. Beccuti, G. Franceschinis1 and S. Haddad (2007) Markov Decision Petri Net and Markov Decision Well-formed Net formalisms Paper presented at the Fourth International Conference on Quantitative Evaluation of Systems, 17–19 September 2007